GED

Math Exercise Book

Review of Essential Skills and Concepts

with 2 GED Math Practice Tests

By

Elise Baniam & Michael Smith

GED Math Exercise Book

Published in the United State of America By

The Math Notion

Email: info@mathnotion.com

Web: www.mathnotion.com

ISBN: 978-1-63620-118-4

About the Author

Elise Baniam has been a math instructor for over a decade now. She graduated in Mathematics. Since 2006, Elise has devoted his time to both teaching and developing exceptional math learning materials. As a Math instructor and test prep expert, Elise has worked with thousands of students. She has used the feedback of her students to develop a unique study program that can be used by students to drastically improve their math score fast and effectively.

– **HiSET Math Workbook**

– **TASC Math Workbook**

– **ASVAB Math Workbook**

– **AFOQT Math Workbook**

–**many Math Education Workbooks**

– **and some Mathematics books …**

As an experienced Math teacher, Mrs. Baniam employs a variety of formats to help students achieve their goals: she teaches students in large groups, and she provides training materials and textbooks through her website and through Amazon.

You can contact Elise via email at:

Elise@mathnotion.com

Get All the Math Prep You Need to Ace the GED Test!

Studying for a test is much easier when you know what will be on it, particularly when you can crack it down into apparent parts. You can then study each section independently.

GED Math Exercise Book helps you achieve the next level of professional achievement. It contains over 2,500 practice problems covering every topic tested on the GED math, making it a critical resource for students to provide them with comprehensive practice. So that you can not only pass the GED Test but earn an advanced score.

Upgraded by our professional instructors, the problems are sensibly categorized into practice sets and reflect those found on the GED in content, form, and style. Students can build fundamental skills in math through targeted practice while easy-to-follow explanations help cement their understanding of the concepts assessed on the GED.

This user-friendly resource includes simple explanations:

- Hands-on experience with all GED math questions.
- Focusing your study time on what is most important.
- Everything you need to know for a High Score.
- Complete review to help you master different concepts.
- These reviews go into detail to cover all math topics on the GED test.
- Hundreds of realistic questions and drills, including new practice questions.
- **2 full-length practice tests** with detailed answer explanations

Effective exercises to help you avoid traps and pacing yourself beat the GED test. It is packed with everything you need to do your best on the test and move toward your graduation.

WWW.MATHNOTION.COM

… So Much More Online!

✓ FREE Math Lessons

✓ More Math Learning Books!

✓ Mathematics Worksheets

✓ Online Math Tutors

For a PDF Version of This Book

SCAN ME

Please Visit www.mathnotion.com

Contents

Chapter 1:

Whole Numbers

Add and Subtract Integers

Find the sum or difference.

1) $(+142) + (-88) =$

2) $(+64) + (-32) =$

3) $288 - 185 =$

4) $(-214) + 157 =$

5) $(-72) + 425 =$

6) $182 + (-265) =$

7) $(-15) + 38 =$

8) $415 - 310 =$

9) $(-18) - (-77) =$

10) $(-88) + (-57) =$

11) $(-124) - 304 =$

12) $1,520 - (-157) =$

13) $24 + (-20) + (-45) + (-20) =$

14) $(-27) + (-24) + 52 + 12 =$

15) $(-5) - 7 + 38 - 21 =$

16) $8 + (-19) + (-29 - 21) =$

17) $(+145) + (+28) + (-157) =$

18) $(-42) + (-32) =$

19) $-14 - 18 - 9 - 31 =$

20) $9 + (-28) =$

21) $134 - 90 - 53 - (-42) =$

22) $(+37) - (-9) =$

23) $(+17) - (+21) - (-19) =$

24) $(+37) - (+9) - (-42) =$

Multiplication and Division

Calculate.

1) $240 \times 7 =$

2) $130 \times 20 =$

3) $(-5) \times 7 \times (-4) =$

4) $-7 \times (-6) \times (-6) =$

5) $11 \times (-11) =$

6) $80 \times (-4) =$

7) $2 \times (-6) \times 8 =$

8) $(-300) \times (-40) =$

9) $(-30) \times (-20) \times 2 =$

10) $115 \times 5 =$

11) $142 \times 50 =$

12) $364 \div 14 =$

13) $(-4,125) \div 5 =$

14) $(-28) \div (-7) =$

15) $288 \div (-18) =$

16) $3,500 \div 28 =$

17) $(-126) \div 3 =$

18) $4,128 \div 4 =$

19) $1,260 \div (-35) =$

20) $3,360 \div 4 =$

21) $(-58) \div 2 =$

22) $(-10,000) \div (-50) =$

23) $0 \div 670 =$

24) $(-1,020) \div 6 =$

25) $5,868 \div 652 =$

26) $(-2,520) \div 4 =$

27) $10,902 \div 3 =$

28) $(-60) \div (-2) =$

Absolute Value

Simplify each equation below.

1) $|-20| =$

2) $-10 + |-40| + 28 =$

3) $|-48| - |-42| + 36 =$

4) $|-8 + 7 - 4| + |5 + 5| =$

5) $3|3 - 9| + 18 =$

6) $|-18| + |-15| =$

7) $|-42 + 18| + 12 - 5 =$

8) $|-16| - |-28| - 7 =$

9) $|-45| - |-4| + 9 =$

10) $|24| - 22 + |-15| =$

11) $\frac{6|4-8|}{8} =$

12) $|-26 + 11| =$

13) $|-24| \times |3| + 10 =$

14) $|-5| + |-21| + 4 - 6 =$

15) $|-6| + |-18| - 29 =$

16) $14 + |-28 + 12| + |-15| =$

17) $26 - |-56| + 20 =$

18) $\frac{|147|}{|7|} + 9 =$

19) $|-6 + 10| + |38 - 18| + 2 =$

20) $|-30 + 18| + |-15| + 10 =$

21) $\frac{|-63|}{7} \times |-9| =$

22) $\frac{12|3 \times 5|}{36} \times \frac{3|-12|}{5} =$

23) $\frac{|7 \times 5|}{35} \times 16 =$

24) $|-24 + 8| \times \frac{|-2 \times 6|}{8} =$

25) $|-130 + 10| - 16 + 16 =$

26) $|-100 + 60| - 40 =$

Ordering Integers and Numbers

Order each set of integers from least to greatest.

1) $16, -10, -5, -1, 2$ ___, ___, ___, ___, ___, ___

2) $-14, -8, 15, 4, 11$ ___, ___, ___, ___, ___, ___

3) $39, -28, -16, 37, -21$ ___, ___, ___, ___, ___, ___

4) $-15, -55, 35, -27, 48$ ___, ___, ___, ___, ___, ___

5) $47, -32, 42, -35, 28$ ___, ___, ___, ___, ___, ___

6) $85, 38, -59, 95, -24$ ___, ___, ___, ___, ___, ___

Order each set of integers from greatest to least.

7) $16, 29, -21, -25, -4$ ___, ___, ___, ___, ___, ___

8) $12, 36, -54, -26, 71$ ___, ___, ___, ___, ___, ___

9) $55, -46, -19, 37, -17$ ___, ___, ___, ___, ___, ___

10) $37, 95, -46, -22, 87$ ___, ___, ___, ___, ___, ___

11) $-9, 79, -65, -78, 84$ ___, ___, ___, ___, ___, ___

12) $-80, -45, -60, 19, 39$ ___, ___, ___, ___, ___, ___

Order of Operations

Evaluate each expression.

1) $8 + (5 \times 2) =$

2) $25 - (6 \times 3) =$

3) $(15 \times 4) + 15 =$

4) $(21 - 6) - (5 \times 4) =$

5) $32 + (30 \div 5) =$

6) $(36 \times 8) \div 12 =$

7) $(63 \div 7) \times (-3) =$

8) $(7 \times 8) + (34 - 18) =$

9) $80 + (3 \times 3) + 5 =$

10) $(20 \times 8) \div (4 + 4) =$

11) $(-10) + (12 \times 5) + 14 =$

12) $(5 \times 7) - (35 \div 5) =$

13) $(7 \times 30 \div 10) - (17 + 13) =$

14) $(14 + 6 - 16) \times 8 - 12 =$

15) $(36 - 18 + 30) \times (96 \div 8) =$

16) $24 + \left(14 - (36 \div 6)\right) =$

17) $(7 + 10 - 4 - 9) + (24 \div 3) =$

18) $(90 - 15) + (16 - 18 + 8) =$

19) $(30 \times 3) + (16 \times 4) - 80 =$

20) $11 + 16 - (21 \times 3) + 25 =$

Factoring

Factor, write prime if prime.

1) 15

2) 41

3) 32

4) 85

5) 62

6) 65

7) 38

8) 10

9) 54

10) 121

11) 45

12) 90

13) 24

14) 33

15) 92

16) 57

17) 86

18) 40

19) 105

20) 80

21) 95

22) 81

23) 126

24) 110

25) 34

26) 98

27) 72

28) 104

Great Common Factor (GCF)

Find the GCF of the numbers.

1) 2, 18

2) 36, 23

3) 45, 35

4) 20, 32

5) 36, 64

6) 32, 42

7) 60, 25

8) 90, 35

9) 72, 9

10) 45, 54

11) 66, 54

12) 35, 70

13) 140, 40

14) 32, 82

15) 48, 96

16) 30, 85

17) 16, 24

18) 80, 100, 40

19) 81, 112

20) 56, 88

21) 20, 10, 50

22) 2, 9, 12

23) 30, 90, 120

24) 51, 33

Least Common Multiple (LCM)

Find the LCM of each.

1) 16, 20

2) 64, 32

3) 10, 20, 30

4) 28, 42

5) 10, 2, 15

6) 25, 5

7) 24, 120, 48

8) 15, 18

9) 13, 26, 54

10) 28, 35

11) 27, 54

12) 220, 44

13) 60, 30, 120

14) 36, 126

15) 40, 8, 5

16) 27, 6

17) 38, 19

18) 22, 44

19) 25, 60

20) 16, 48

21) 34, 20

22) 24, 28

23) 70, 140

24) 63, 18

Divisibility Rule

Apply the divisibility rules to find the factors of each number.

1) 16   2, 3, 4, 5, 6, 9, 10   13) 38   2, 3, 4, 5, 6, 9, 10

2) 121   2, 3, 4, 5, 6, 9, 10   14) 185   2, 3, 4, 5, 6, 9, 10

3) 252   2, 3, 4, 5, 6, 9, 10   15) 905   2, 3, 4, 5, 6, 9, 10

4) 74   2, 3, 4, 5, 6, 9, 10   16) 157   2, 3, 4, 5, 6, 9, 10

5) 241   2, 3, 4, 5, 6, 9, 10   17) 540   2, 3, 4, 5, 6, 9, 10

6) 155   2, 3, 4, 5, 6, 9, 10   18) 340   2, 3, 4, 5, 6, 9, 10

7) 65   2, 3, 4, 5, 6, 9, 10   19) 480   2, 3, 4, 5, 6, 9, 10

8) 320   2, 3, 4, 5, 6, 9, 10   20) 2,750   2, 3, 4, 5, 6, 9, 10

9) 1,134   2, 3, 4, 5, 6, 9, 10   21) 330   2, 3, 4, 5, 6, 9, 10

10) 68   2, 3, 4, 5, 6, 9, 10   22) 346   2, 3, 4, 5, 6, 9, 10

11) 754   2, 3, 4, 5, 6, 9, 10   23) 108   2, 3, 4, 5, 6, 9, 10

12) 128   2, 3, 4, 5, 6, 9, 10   24) 656   2, 3, 4, 5, 6, 9, 10

Answers of Worksheets

Add and Subtract Integers

1) 54	9) 59	17) 16
2) 32	10) −145	18) −74
3) 103	11) −428	19) −72
4) −57	12) 1,677	20) −19
5) 353	13) −61	21) 33
6) −83	14) 13	22) 46
7) 23	15) 5	23) 15
8) 105	16) −61	24) 50

Multiplication and Division

1) 1,680	11) 7,100	21) −29
2) 2,600	12) 26	22) 200
3) 140	13) −825	23) 0
4) −252	14) 4	24) −170
5) −121	15) −16	25) 9
6) −320	16) 125	26) −630
7) −96	17) −42	27) 3,634
8) 12,000	18) 1,032	28) 30
9) 1,200	19) −36	
10) 575	20) 840	

Absolute Value

1) 20	8) −19	15) −5
2) 58	9) 50	16) 45
3) 42	10) 17	17) −10
4) 15	11) 3	18) 30
5) 36	12) 15	19) 26
6) 33	13) 82	20) 37
7) 31	14) 24	21) 81

22) 36   24) 24   26) 0

23) 16   25) 120

Ordering Integers and Numbers

1) -10, −5, −1, 2, 16   7) 29, 16, −4, −21, −25

2) −14, −8, 4, 11, 15   8) 71, 36, 12, −26, −54

3) −28, −21, −16, 37, 39   9) 55, 37, −17, −19, −46

4) −55, −27, −15, 35, 48   10) 95, 87, 37, −22, −46

5) −35, −32, 28, 42, 47   11) 84, 79, −9, −65, −78

6) −59, −24, 38, 85, 95   12) 39, 19, −45, −60, −80

Order of Operations

1) 18   6) 24   11) 64   16) 32

2) 7   7) −27   12) 28   17) 12

3) 75   8) 72   13) −9   18) 81

4) −5   9) 94   14) 20   19) 74

5) 38   10) 20   15) 576   20) −11

Factoring

1) 1, 3, 5, 15   15) 1, 2, 4, 23, 46, 92

2) 1, 41   16) 1, 3, 19, 57

3) 1, 2, 4, 8, 16, 32   17) 1, 2, 43, 86

4) 1, 5, 17, 85   18) 1, 2, 4, 5, 8, 10, 20, 40

5) 1, 2, 31, 62   19) 1, 3, 5, 7, 15, 21, 35, 105

6) 1, 5, 13, 65   20) 1, 2, 4, 5, 8, 10, 16, 20, 40, 80

7) 1, 2, 19, 38   21) 1, 3, 9, 27, 81

8) 1, 2, 5, 10   22) 1, 2, 3, 4, 6, 9, 12, 18, 36

9) 1, 2, 3, 6, 9, 18, 27, 54   23) 1, 2, 3, 6, 7, 9, 14, 18, 21, 42, 63, 126

10) 1, 11, 121   24) 1, 2, 5, 10, 11, 22, 55, 110

11) 1, 3, 5, 9, 15, 45   25) 1, 2, 17, 34

12) 1, 2, 3, 5, 6, 9, 10, 15, 18, 30, 45, 90   26) 1, 2, 7, 14, 49, 98

13) 1, 2, 3, 4, 6, 8, 12, 24   27) 1, 2, 3, 4, 6, 8, 12, 18, 24, 36, 72

14) 1, 3, 11, 33   28) 1, 2, 4, 8, 13, 26, 52, 104

Great Common Factor (GCF)

1) 2	9) 9	17) 8
2) 1	10) 9	18) 20
3) 5	11) 6	19) 1
4) 5	12) 35	20) 8
5) 4	13) 20	21) 10
6) 2	14) 2	22) 1
7) 5	15) 48	23) 30
8) 5	16) 5	24) 3

Least Common Multiple (LCM)

1) 80	9) 702	17) 38
2) 64	10) 140	18) 44
3) 60	11) 54	19) 300
4) 84	12) 220	20) 48
5) 30	13) 120	21) 340
6) 25	14) 252	22) 168
7) 240	15) 40	23) 140
8) 90	16) 54	24) 126

Divisibility Rule

1) 16	$\underline{2}$, 3, $\underline{4}$, 5, 6, 9, 10	13) 38	$\underline{2}$, 3, 4, 5, 6, 9, 10
2) 121	2, 3, 4, 5, 6, 9, 10	14) 185	2, 3, 4, $\underline{5}$, 6, 9, 10
3) 252	$\underline{2}$, $\underline{3}$, $\underline{4}$, 5, $\underline{6}$, 9, 10	15) 905	2, 3, 4, $\underline{5}$, 6, 9, 10
4) 74	$\underline{2}$, 3, 4, 5, 6, 9, 10	16) 157	2, 3, 4, 5, 6, 9, 10
5) 241	2, 3, 4, 5, 6, 9, 10	17) 540	$\underline{2}$, $\underline{3}$, $\underline{4}$, $\underline{5}$, $\underline{6}$, $\underline{9}$, $\underline{10}$
6) 155	2, 3, 4, $\underline{5}$, 6, 9, 10	18) 340	$\underline{2}$, 3, $\underline{4}$, $\underline{5}$, 6, 9, $\underline{10}$
7) 65	2, 3, 4, $\underline{5}$, 6, 9, 10	19) 480	$\underline{2}$, $\underline{3}$, $\underline{4}$, $\underline{5}$, $\underline{6}$, 9, $\underline{10}$
8) 320	2, 3, $\underline{4}$, $\underline{5}$, 6, 9, 10	20) 2,750	$\underline{2}$, 3, 4, $\underline{5}$, 6, 9, $\underline{10}$
9) 1,134	$\underline{2}$, $\underline{3}$, 4, 5, $\underline{6}$, $\underline{9}$, 10	21) 330	$\underline{2}$, $\underline{3}$, 4, $\underline{5}$, $\underline{6}$, 9, $\underline{10}$
10) 68	$\underline{2}$, 3, $\underline{4}$, 5, 6, 9, 10	22) 346	$\underline{2}$, 3, 4, 5, 6, 9, 10
11) 754	$\underline{2}$, 3, 4, 5, 6, 9, 10	23) 108	$\underline{2}$, $\underline{3}$, $\underline{4}$, 5, $\underline{6}$, 9, 10
12) 128	$\underline{2}$, 3, $\underline{4}$, 5, 6, 9, 10	24) 656	$\underline{2}$, 3, $\underline{4}$, 5, 6, 9, 10

Chapter 2:

Fractions

Adding Fractions – Like Denominator

Find each sum.

1) $\dfrac{1}{7} + \dfrac{2}{7} =$

2) $\dfrac{2}{9} + \dfrac{1}{9} =$

3) $\dfrac{1}{11} + \dfrac{2}{11} =$

4) $\dfrac{7}{17} + \dfrac{1}{17} =$

5) $\dfrac{4}{23} + \dfrac{1}{23} =$

6) $\dfrac{5}{51} + \dfrac{6}{51} =$

7) $\dfrac{2}{4} + \dfrac{11}{4} =$

8) $\dfrac{1}{19} + \dfrac{3}{19} =$

9) $\dfrac{3}{91} + \dfrac{6}{91} =$

10) $\dfrac{1}{13} + \dfrac{1}{13} =$

11) $\dfrac{1}{81} + \dfrac{1}{81} =$

12) $\dfrac{4}{17} + \dfrac{6}{17} =$

13) $\dfrac{2}{20} + \dfrac{17}{20} =$

14) $\dfrac{4}{25} + \dfrac{7}{25} =$

15) $\dfrac{6}{14} + \dfrac{3}{14} =$

16) $\dfrac{12}{30} + \dfrac{5}{30} =$

17) $\dfrac{1}{31} + \dfrac{1}{31} =$

18) $\dfrac{29}{7} + \dfrac{3}{7} =$

19) $\dfrac{18}{11} + \dfrac{5}{11} =$

20) $\dfrac{25}{73} + \dfrac{11}{73} =$

Adding Fractions – Unlike Denominator

Add the fractions and simplify the answers.

1) $\frac{1}{8} + \frac{2}{5} =$

2) $\frac{3}{7} + \frac{1}{2} =$

3) $\frac{1}{4} + \frac{2}{9} =$

4) $\frac{3}{5} + \frac{1}{2} =$

5) $\frac{7}{18} + \frac{1}{3} =$

6) $\frac{13}{54} + \frac{5}{18} =$

7) $\frac{5}{8} + \frac{1}{6} =$

8) $\frac{3}{10} + \frac{1}{4} =$

9) $\frac{5}{11} + \frac{2}{4} =$

10) $\frac{1}{9} + \frac{4}{7} =$

11) $\frac{5}{18} + \frac{3}{8} =$

12) $\frac{7}{32} + \frac{3}{4} =$

13) $\frac{18}{130} + \frac{4}{10} =$

14) $\frac{8}{63} + \frac{3}{7} =$

15) $\frac{11}{64} + \frac{1}{4} =$

16) $\frac{4}{15} + \frac{2}{5} =$

17) $\frac{4}{7} + \frac{3}{6} =$

18) $\frac{5}{72} + \frac{2}{9} =$

19) $\frac{2}{15} + \frac{1}{25} =$

20) $\frac{5}{12} + \frac{3}{8} =$

21) $\frac{7}{88} + \frac{1}{8} =$

22) $\frac{7}{12} + \frac{2}{5} =$

23) $\frac{3}{72} + \frac{1}{4} =$

24) $\frac{2}{27} + \frac{1}{18} =$

Subtracting Fractions – Like Denominator

Find the difference.

1) $\frac{8}{7} - \frac{1}{7} =$

2) $\frac{7}{20} - \frac{3}{20} =$

3) $\frac{11}{21} - \frac{8}{21} =$

4) $\frac{13}{8} - \frac{7}{8} =$

5) $\frac{15}{16} - \frac{13}{16} =$

6) $\frac{18}{31} - \frac{10}{31} =$

7) $\frac{8}{25} - \frac{2}{25} =$

8) $\frac{17}{27} - \frac{2}{27} =$

9) $\frac{7}{10} - \frac{3}{10} =$

10) $\frac{24}{35} - \frac{4}{35} =$

11) $\frac{11}{5} - \frac{3}{5} =$

12) $\frac{28}{56} - \frac{18}{56} =$

13) $\frac{10}{12} - \frac{2}{12} =$

14) $\frac{22}{43} - \frac{11}{43} =$

15) $\frac{4}{7} - \frac{3}{7} =$

16) $\frac{18}{29} - \frac{15}{29} =$

17) $\frac{4}{5} - \frac{3}{5} =$

18) $\frac{42}{53} - \frac{38}{53} =$

19) $\frac{8}{31} - \frac{3}{31} =$

20) $\frac{32}{19} - \frac{30}{19} =$

21) $\frac{9}{34} - \frac{5}{34} =$

22) $\frac{28}{62} - \frac{17}{62} =$

23) $\frac{23}{42} - \frac{19}{42} =$

24) $\frac{33}{83} - \frac{21}{83} =$

Subtracting Fractions – Unlike Denominator

Solve each problem.

1) $\frac{3}{5} - \frac{1}{6} =$

2) $\frac{5}{6} - \frac{1}{8} =$

3) $\frac{7}{6} - \frac{3}{11} =$

4) $\frac{5}{7} - \frac{4}{15} =$

5) $\frac{6}{7} - \frac{3}{14} =$

6) $\frac{7}{12} - \frac{7}{18} =$

7) $\frac{17}{20} - \frac{2}{5} =$

8) $\frac{2}{3} - \frac{1}{16} =$

9) $\frac{6}{7} - \frac{4}{9} =$

10) $\frac{3}{8} - \frac{5}{32} =$

11) $\frac{3}{5} - \frac{7}{40} =$

12) $\frac{5}{6} - \frac{7}{30} =$

13) $\frac{6}{7} - \frac{4}{21} =$

14) $\frac{5}{3} - \frac{8}{15} =$

15) $\frac{2}{11} - \frac{3}{22} =$

16) $\frac{5}{6} - \frac{4}{54} =$

17) $\frac{7}{24} - \frac{7}{32} =$

18) $\frac{3}{4} - \frac{3}{5} =$

19) $\frac{1}{2} - \frac{2}{9} =$

20) $\frac{5}{11} - \frac{3}{13} =$

Converting Mix Numbers

Convert the following mixed numbers into improper fractions.

1) $5\frac{5}{4} =$

2) $5\frac{7}{12} =$

3) $4\frac{1}{7} =$

4) $6\frac{1}{10} =$

5) $3\frac{1}{4} =$

6) $3\frac{19}{21} =$

7) $5\frac{9}{10} =$

8) $4\frac{7}{12} =$

9) $3\frac{10}{11} =$

10) $6\frac{2}{5} =$

11) $8\frac{2}{3} =$

12) $3\frac{7}{16} =$

13) $6\frac{8}{13} =$

14) $4\frac{8}{11} =$

15) $7\frac{1}{4} =$

16) $5\frac{6}{11} =$

17) $8\frac{1}{5} =$

18) $3\frac{7}{12} =$

19) $6\frac{1}{22} =$

20) $3\frac{2}{3} =$

21) $7\frac{1}{19} =$

22) $4\frac{7}{11} =$

23) $1\frac{5}{8} =$

24) $9\frac{5}{17} =$

Converting improper Fractions

Convert the following improper fractions into mixed numbers

1) $\frac{68}{19} =$

2) $\frac{79}{33} =$

3) $\frac{49}{17} =$

4) $\frac{56}{23} =$

5) $\frac{79}{18} =$

6) $\frac{137}{42} =$

7) $\frac{120}{33} =$

8) $\frac{26}{5} =$

9) $\frac{33}{19} =$

10) $\frac{13}{2} =$

11) $\frac{39}{4} =$

12) $\frac{161}{50} =$

13) $\frac{89}{77} =$

14) $\frac{42}{19} =$

15) $\frac{110}{13} =$

16) $\frac{65}{4} =$

17) $\frac{122}{9} =$

18) $\frac{81}{16} =$

19) $\frac{37}{6} =$

20) $\frac{67}{22} =$

21) $\frac{5}{4} =$

22) $\frac{79}{13} =$

23) $\frac{51}{11} =$

24) $\frac{36}{5} =$

Addition Mix Numbers

Add the following fractions.

1) $2\frac{1}{5} + 3\frac{2}{5} =$

2) $5\frac{3}{11} + 3\frac{4}{11} =$

3) $2\frac{2}{10} + 3\frac{1}{10} =$

4) $3\frac{3}{16} + 2\frac{1}{4} =$

5) $2\frac{3}{7} + 3\frac{4}{21} =$

6) $6\frac{2}{7} + 3\frac{1}{2} =$

7) $2\frac{8}{27} + 2\frac{2}{18} =$

8) $2\frac{3}{4} + 3\frac{1}{3} =$

9) $4\frac{5}{6} + 1\frac{1}{6} =$

10) $3\frac{5}{7} + 1\frac{3}{7} =$

11) $4\frac{1}{2} + 2\frac{2}{5} =$

12) $5\frac{1}{4} + 2\frac{5}{6} =$

13) $4\frac{1}{7} + 2\frac{2}{7} =$

14) $6\frac{5}{9} + 4\frac{2}{18} =$

15) $6\frac{3}{7} + 3\frac{1}{2} =$

16) $5\frac{2}{3} + 1\frac{4}{7} =$

17) $4\frac{5}{6} + 6\frac{1}{4} =$

18) $2\frac{2}{5} + 3\frac{3}{8} =$

19) $3\frac{1}{6} + 2\frac{4}{9} =$

20) $5\frac{3}{5} + 7\frac{2}{7} =$

21) $4\frac{5}{8} + 1\frac{1}{5} =$

22) $6\frac{1}{7} + 4\frac{4}{13} =$

23) $2\frac{7}{11} + 2\frac{4}{5} =$

24) $3\frac{1}{6} + 1\frac{5}{12} =$

Subtracting Mix Numbers

Subtract the following fractions.

1) $5 \frac{1}{7} - 3 \frac{1}{7} =$

2) $8 \frac{5}{16} - 8 \frac{2}{16} =$

3) $8 \frac{5}{36} - 7 \frac{1}{36} =$

4) $4 \frac{1}{20} - 1 \frac{1}{15} =$

5) $3 \frac{1}{3} - 2 \frac{1}{6} =$

6) $8 \frac{1}{2} - 3 \frac{2}{5} =$

7) $7 \frac{5}{8} - 3 \frac{3}{8} =$

8) $9 \frac{9}{13} - 4 \frac{6}{13} =$

9) $5 \frac{7}{12} - 2 \frac{5}{12} =$

10) $4 \frac{4}{7} - 1 \frac{3}{7} =$

11) $7 \frac{1}{5} - 2 \frac{1}{10} =$

12) $4 \frac{5}{6} - 2 \frac{1}{6} =$

13) $3 \frac{2}{40} - 1 \frac{1}{5} =$

14) $4 \frac{1}{8} - 2 \frac{1}{16} =$

15) $14 \frac{4}{15} - 11 \frac{2}{15} =$

16) $6 \frac{2}{4} - 1 \frac{1}{4} =$

17) $4 \frac{1}{7} - 2 \frac{3}{7} =$

18) $5 \frac{1}{16} - 2 \frac{1}{4} =$

19) $6 \frac{2}{3} - 1 \frac{1}{9} =$

20) $4 \frac{3}{25} - 4 \frac{1}{75} =$

21) $9 \frac{9}{22} - 5 \frac{1}{4} =$

22) $8 \frac{4}{10} - 2 \frac{3}{40} =$

23) $3 \frac{2}{6} - 2 \frac{1}{18} =$

24) $7 \frac{9}{13} - 3 \frac{3}{13} =$

Simplify Fractions

Reduce these fractions to lowest terms

1) $\frac{18}{12} =$

2) $\frac{10}{15} =$

3) $\frac{32}{40} =$

4) $\frac{27}{36} =$

5) $\frac{6}{36} =$

6) $\frac{27}{63} =$

7) $\frac{16}{28} =$

8) $\frac{48}{60} =$

9) $\frac{8}{72} =$

10) $\frac{30}{12} =$

11) $\frac{45}{60} =$

12) $\frac{30}{90} =$

13) $\frac{21}{35} =$

14) $\frac{7}{28} =$

15) $\frac{24}{84} =$

16) $\frac{34}{51} =$

17) $\frac{66}{55} =$

18) $\frac{36}{135} =$

19) $\frac{21}{56} =$

20) $\frac{64}{56} =$

21) $\frac{140}{280} =$

22) $\frac{138}{731} =$

23) $\frac{175}{35} =$

24) $\frac{170}{680} =$

Multiplying Fractions

Find the product.

1) $\dfrac{8}{5} \times \dfrac{2}{12} =$

2) $\dfrac{8}{44} \times \dfrac{10}{16} =$

3) $\dfrac{8}{30} \times \dfrac{12}{16} =$

4) $\dfrac{9}{14} \times \dfrac{21}{36} =$

5) $\dfrac{14}{15} \times \dfrac{5}{7} =$

6) $\dfrac{16}{19} \times \dfrac{3}{4} =$

7) $\dfrac{4}{9} \times \dfrac{9}{8} =$

8) $\dfrac{87}{63} \times 0 =$

9) $\dfrac{5}{16} \times \dfrac{32}{14} =$

10) $\dfrac{32}{45} \times \dfrac{5}{8} =$

11) $\dfrac{34}{26} \times \dfrac{13}{17} =$

12) $\dfrac{6}{42} \times \dfrac{7}{36} =$

13) $\dfrac{26}{16} \times \dfrac{24}{8} =$

14) $\dfrac{10}{27} \times \dfrac{18}{5} =$

15) $\dfrac{30}{54} \times \dfrac{16}{6} =$

16) $\dfrac{24}{14} \times 7 =$

17) $\dfrac{10}{33} \times \dfrac{66}{35} =$

18) $\dfrac{10}{18} \times \dfrac{9}{20} =$

19) $\dfrac{7}{11} \times \dfrac{8}{21} =$

20) $\dfrac{26}{24} \times \dfrac{8}{52} =$

21) $\dfrac{30}{15} \times \dfrac{1}{60} =$

22) $\dfrac{20}{27} \times \dfrac{18}{100} =$

23) $\dfrac{8}{21} \times \dfrac{7}{64} =$

24) $\dfrac{100}{200} \times \dfrac{600}{800} =$

Multiplying Mixed Number

Multiply. Reduce to lowest terms.

1) $2\frac{6}{10} \times 1\frac{6}{8} =$

2) $1\frac{15}{12} \times 1\frac{2}{9} =$

3) $2\frac{3}{7} \times 1\frac{2}{9} =$

4) $3\frac{1}{7} \times 2\frac{1}{2} =$

5) $4\frac{3}{4} \times 1\frac{1}{4} =$

6) $3\frac{1}{2} \times 1\frac{4}{5} =$

7) $3\frac{3}{4} \times 1\frac{1}{2} =$

8) $5\frac{2}{3} \times 3\frac{1}{3} =$

9) $3\frac{2}{3} \times 3\frac{1}{2} =$

10) $2\frac{1}{3} \times 3\frac{1}{2} =$

11) $4\frac{3}{4} \times 3\frac{2}{3} =$

12) $2\frac{4}{11} \times 2\frac{1}{7} =$

13) $2\frac{2}{7} \times 1\frac{1}{5} =$

14) $3\frac{1}{3} \times 1\frac{1}{5} =$

15) $2\frac{2}{3} \times 3\frac{1}{2} =$

16) $2\frac{1}{8} \times 2\frac{2}{5} =$

17) $2\frac{1}{4} \times 1\frac{2}{3} =$

18) $2\frac{3}{5} \times 1\frac{1}{4} =$

19) $2\frac{3}{5} \times 1\frac{5}{8} =$

20) $3\frac{1}{6} \times 2\frac{5}{7} =$

21) $2\frac{5}{8} \times 1\frac{1}{5} =$

22) $2\frac{5}{7} \times 3\frac{1}{6} =$

Dividing Fractions

Divide these fractions.

1) $1 \div \frac{1}{8} =$

2) $\frac{9}{17} \div 9 =$

3) $\frac{11}{20} \div \frac{5}{11} =$

4) $\frac{25}{60} \div \frac{5}{4} =$

5) $\frac{6}{23} \div \frac{4}{23} =$

6) $\frac{4}{16} \div \frac{18}{24} =$

7) $0 \div \frac{1}{9} =$

8) $\frac{12}{16} \div \frac{8}{9} =$

9) $\frac{8}{12} \div \frac{4}{18} =$

10) $\frac{9}{14} \div \frac{3}{7} =$

11) $\frac{8}{15} \div \frac{25}{16} =$

12) $\frac{35}{16} \div \frac{15}{8} =$

13) $\frac{11}{15} \div \frac{11}{5} =$

14) $\frac{8}{16} \div \frac{20}{6} =$

15) $\frac{40}{24} \div \frac{48}{80} =$

16) $\frac{7}{30} \div \frac{63}{5} =$

17) $\frac{36}{8} \div \frac{18}{24} =$

18) $9 \div \frac{1}{2} =$

19) $\frac{48}{35} \div \frac{8}{7} =$

20) $\frac{3}{36} \div \frac{9}{6} =$

21) $\frac{4}{7} \div \frac{12}{14} =$

22) $\frac{8}{40} \div \frac{10}{5} =$

Dividing Mixed Number

Divide the following mixed numbers. Cancel and simplify when possible.

1) $4\frac{1}{6} \div 4\frac{1}{5} =$

2) $3\frac{1}{8} \div 1\frac{1}{4} =$

3) $3\frac{1}{4} \div 2\frac{2}{7} =$

4) $4\frac{1}{3} \div 4\frac{1}{2} =$

5) $3\frac{1}{7} \div 1\frac{2}{5} =$

6) $3\frac{3}{5} \div 2\frac{2}{6} =$

7) $4\frac{3}{5} \div 2\frac{1}{3} =$

8) $2\frac{4}{9} \div 1\frac{1}{9} =$

9) $3\frac{5}{6} \div 3\frac{1}{2} =$

10) $9\frac{1}{9} \div 3\frac{2}{3} =$

11) $2\frac{2}{7} \div 4\frac{1}{7} =$

12) $4\frac{3}{8} \div 1\frac{3}{4} =$

13) $5\frac{1}{8} \div 1\frac{1}{12} =$

14) $6\frac{3}{8} \div 3\frac{1}{3} =$

15) $4\frac{2}{5} \div 1\frac{1}{5} =$

16) $2\frac{1}{2} \div 2\frac{2}{9} =$

17) $7\frac{1}{6} \div 5\frac{3}{8} =$

18) $5\frac{1}{2} \div 4\frac{1}{3} =$

19) $4\frac{5}{7} \div 1\frac{1}{3} =$

20) $3\frac{5}{6} \div 1\frac{1}{4} =$

21) $8\frac{1}{3} \div 5\frac{1}{4} =$

22) $3\frac{1}{11} \div 1\frac{1}{5} =$

23) $4\frac{1}{6} \div 5\frac{5}{6} =$

24) $2\frac{1}{14} \div 2\frac{1}{7} =$

Comparing Fractions

Compare the fractions, and write >, < or =

1) $\dfrac{28}{3}$ ____ $\dfrac{48}{15}$

2) $\dfrac{96}{3}$ ____ $\dfrac{14}{5}$

3) $\dfrac{8}{9}$ ____ $\dfrac{6}{4}$

4) $\dfrac{12}{4}$ ____ $\dfrac{13}{9}$

5) $\dfrac{1}{8}$ ____ $\dfrac{2}{3}$

6) $\dfrac{10}{6}$ ____ $\dfrac{16}{7}$

7) $\dfrac{12}{13}$ ____ $\dfrac{7}{9}$

8) $\dfrac{20}{14}$ ____ $\dfrac{25}{3}$

9) $4\dfrac{1}{12}$ ____ $6\dfrac{1}{3}$

10) $8\dfrac{1}{6}$ ____ $3\dfrac{1}{8}$

11) $3\dfrac{1}{2}$ ____ $3\dfrac{1}{5}$

12) $7\dfrac{5}{8}$ ____ $7\dfrac{2}{9}$

13) $3\dfrac{4}{16}$ ____ $5\dfrac{6}{10}$

14) $\dfrac{1}{15}$ ____ $\dfrac{3}{7}$

15) $\dfrac{31}{25}$ ____ $\dfrac{19}{83}$

16) $\dfrac{12}{100}$ ____ $\dfrac{6}{62}$

17) $15\dfrac{1}{4}$ ____ $15\dfrac{1}{9}$

18) $\dfrac{1}{5}$ ____ $\dfrac{1}{9}$

19) $\dfrac{1}{7}$ ____ $\dfrac{1}{13}$

20) $\dfrac{1}{18}$ ____ $\dfrac{8}{15}$

21) $\dfrac{7}{22}$ ____ $\dfrac{9}{76}$

22) $\dfrac{4}{5}$ ____ $\dfrac{2}{5}$

23) $3\dfrac{7}{6}$ ____ $4\dfrac{2}{12}$

24) $3\dfrac{25}{6}$ ____ $4\dfrac{5}{6}$

Answers of Worksheets

Adding Fractions – Like Denominator

1) $\frac{3}{7}$

2) $\frac{3}{9}$

3) $\frac{3}{11}$

4) $\frac{8}{17}$

5) $\frac{5}{23}$

6) $\frac{11}{51}$

7) $\frac{13}{4}$

8) $\frac{4}{19}$

9) $\frac{9}{91}$

10) $\frac{2}{13}$

11) $\frac{2}{81}$

12) $\frac{10}{17}$

13) $\frac{19}{20}$

14) $\frac{11}{25}$

15) $\frac{9}{14}$

16) $\frac{17}{30}$

17) $\frac{2}{31}$

18) $\frac{32}{7}$

19) $\frac{23}{11}$

20) $\frac{36}{73}$

Adding Fractions – Unlike Denominator

1) $\frac{21}{40}$

2) $\frac{13}{14}$

3) $\frac{17}{36}$

4) $\frac{27}{22}$

5) $\frac{13}{18}$

6) $\frac{14}{27}$

7) $\frac{19}{24}$

8) $\frac{11}{20}$

9) $\frac{21}{22}$

10) $\frac{43}{63}$

11) $\frac{47}{72}$

12) $\frac{31}{32}$

13) $\frac{7}{12}$

14) $\frac{5}{9}$

15) $\frac{27}{64}$

16) $\frac{2}{3}$

17) $\frac{15}{14}$

18) $\frac{7}{24}$

19) $\frac{13}{75}$

20) $\frac{19}{24}$

21) $\frac{9}{44}$

22) $\frac{59}{60}$

23) $\frac{7}{24}$

24) $\frac{7}{54}$

Subtracting Fractions – Like Denominator

1) 1

2) $\frac{1}{5}$

3) $\frac{1}{7}$

4) $\frac{3}{4}$

5) $\frac{1}{8}$

6) $\frac{8}{31}$

7) $\frac{6}{25}$

8) $\frac{5}{9}$

9) $\frac{2}{5}$

10) $\frac{4}{7}$

11) $\frac{8}{5}$

12) $\frac{5}{28}$

13) $\frac{2}{3}$

14) $\frac{11}{43}$

15) $\frac{1}{7}$

16) $\frac{3}{29}$

17) $\frac{1}{5}$

18) $\frac{4}{53}$

19) $\frac{5}{31}$

20) $\frac{2}{19}$

21) $\frac{2}{17}$

22) $\frac{11}{62}$

23) $\frac{2}{21}$

24) $\frac{12}{83}$

Subtracting Fractions – Unlike Denominator

1) $\frac{13}{30}$

2) $\frac{17}{24}$

3) $\frac{59}{66}$

4) $\frac{47}{105}$

5) $\frac{9}{14}$

6) $\frac{7}{36}$

7) $\frac{9}{20}$

8) $\frac{29}{48}$

9) $\frac{26}{63}$

10) $\frac{7}{32}$

11) $\frac{17}{40}$

12) $\frac{3}{5}$

13) $\frac{2}{3}$

14) $\frac{17}{15}$

15) $\frac{1}{22}$

16) $\frac{41}{54}$

17) $\frac{7}{96}$

18) $\frac{3}{20}$

19) $\frac{5}{18}$

20) $\frac{22}{143}$

Converting Mix Numbers

1) $\frac{25}{4}$

2) $\frac{67}{12}$

3) $\frac{29}{7}$

4) $\frac{61}{10}$

5) $\frac{13}{4}$

6) $\frac{82}{21}$

7) $\frac{59}{10}$

8) $\frac{55}{12}$

9) $\frac{43}{11}$

10) $\frac{32}{5}$

11) $\frac{26}{3}$

12) $\frac{55}{16}$

13) $\frac{86}{13}$

14) $\frac{52}{11}$

15) $\frac{29}{4}$

16) $\frac{61}{11}$

17) $\frac{41}{5}$

18) $\frac{43}{12}$

19) $\frac{133}{22}$

20) $\frac{11}{3}$

21) $\frac{134}{19}$

22) $\frac{51}{11}$

23) $\frac{13}{8}$

24) $\frac{158}{17}$

Converting improper Fractions

1) $3\frac{11}{19}$

2) $2\frac{13}{33}$

3) $2\frac{15}{17}$

4) $2\frac{10}{23}$

5) $4\frac{7}{18}$

6) $3\frac{11}{42}$

7) $3\frac{21}{33}$

8) $5\frac{1}{5}$

9) $1\frac{14}{19}$

10) $6\frac{1}{2}$

11) $9\frac{3}{4}$

12) $3\frac{11}{50}$

13) $1\frac{12}{77}$

14) $2\frac{4}{19}$

15) $8\frac{6}{13}$

16) $16\frac{1}{4}$

17) $13\frac{5}{9}$

18) $5\frac{1}{16}$

19) $6\frac{1}{6}$

20) $3\frac{1}{22}$

21) $1\frac{1}{4}$

22) $6\frac{1}{13}$

23) $5\frac{1}{11}$

24) $7\frac{1}{5}$

Adding Mix Numbers

1) $10\frac{3}{5}$

2) $8\frac{7}{11}$

3) $5\frac{3}{10}$

4) $7\frac{5}{16}$

5) $4\frac{19}{21}$

6) $24\frac{5}{14}$

7) $4\frac{11}{27}$

8) $6\frac{1}{12}$

9) 9

10) $5\frac{1}{7}$

11) $6\frac{9}{10}$

12) $8\frac{1}{12}$

13) 8

14) $11\frac{1}{3}$

15) $24\frac{1}{2}$

16) $7\frac{5}{21}$

17) $11\frac{1}{12}$

18) $5\frac{31}{40}$

19) $5\frac{11}{18}$

20) $20\frac{31}{35}$

21) $7\frac{33}{40}$

22) $12\frac{9}{91}$

23) $7\frac{14}{55}$

24) $6\frac{5}{12}$

Subtracting Mix Numbers

1) 2

2) $\frac{3}{16}$

3) $1\frac{1}{9}$

4) $2\frac{59}{60}$

5) $1\frac{1}{6}$

6) $5\frac{1}{10}$

7) $4\frac{1}{4}$

8) $5\frac{3}{13}$

9) $3\frac{1}{6}$

10) $3\frac{1}{7}$

11) $5\frac{1}{10}$

12) $2\frac{2}{3}$

13) $1\frac{17}{20}$

14) $2\frac{1}{16}$

15) $3\frac{2}{15}$

16) $5\frac{1}{4}$

17) $1\frac{5}{7}$

18) $2\frac{13}{16}$

19) $5\frac{5}{9}$

20) $\frac{8}{75}$

21) $4\frac{7}{44}$

22) $6\frac{13}{40}$

23) $1\frac{5}{18}$

24) $4\frac{6}{13}$

Simplify Fractions

1) $\frac{3}{2}$

2) $\frac{2}{3}$

3) $\frac{4}{5}$

4) $\frac{3}{4}$

5) $\frac{1}{6}$

6) $\frac{3}{7}$

7) $\frac{4}{7}$

8) $\frac{4}{5}$

9) $\frac{1}{9}$

10) $\frac{5}{2}$

11) $\frac{3}{4}$

12) $\frac{1}{3}$

13) $\frac{3}{5}$

14) $\frac{1}{4}$

15) $\frac{2}{7}$

16) $\frac{2}{3}$

17) $\frac{6}{5}$

18) $\frac{4}{15}$

19) $\frac{3}{8}$

20) $\frac{8}{7}$

21) $\frac{1}{2}$

22) $\frac{6}{31}$

23) 5

24) $\frac{1}{4}$

Multiplying Fractions

1) $\frac{4}{15}$

2) $\frac{5}{44}$

3) $\frac{1}{5}$

4) $\frac{3}{8}$

5) $\frac{2}{3}$

6) $\frac{12}{19}$

7) $\frac{1}{2}$

8) 0

9) $\frac{5}{7}$

10) $\frac{4}{9}$

11) 1

12) $\frac{1}{36}$

13) $\frac{39}{8}$

14) $\frac{4}{3}$

15) $\frac{40}{27}$

16) 12

17) $\frac{4}{7}$

18) $\frac{1}{4}$

19) $\frac{8}{33}$

20) $\frac{1}{6}$

21) $\frac{1}{30}$

22) $\frac{2}{15}$

23) $\frac{1}{24}$

24) $\frac{3}{8}$

Multiplying Mixed Number

1) $4\frac{11}{20}$

2) $2\frac{3}{4}$

3) $2\frac{61}{63}$

4) $7\frac{6}{7}$

5) $5\frac{15}{16}$

6) $6\frac{3}{10}$

7) $5\frac{5}{8}$

8) $18\frac{8}{9}$

9) $12\frac{5}{6}$

10) $8\frac{1}{6}$

11) $17\frac{5}{12}$

12) $5\frac{5}{77}$

13) $2\frac{26}{35}$

14) 4

15) $9\frac{1}{3}$

16) $5\frac{1}{10}$

17) $3\frac{3}{4}$

18) $3\frac{1}{4}$

19) $4\frac{9}{40}$

20) $8\frac{25}{42}$

21) $3\frac{3}{20}$

22) $8\frac{25}{42}$

Dividing Fractions

1) 8
2) $\frac{1}{17}$
3) $\frac{1}{100}$
4) $\frac{1}{48}$
5) $\frac{3}{2}$
6) $\frac{1}{3}$
7) 0
8) $\frac{27}{32}$

9) 3
10) $\frac{3}{2}$
11) $\frac{128}{375}$
12) $\frac{7}{6}$
13) $\frac{1}{75}$
14) $\frac{1}{240}$
15) 1
16) $\frac{1}{54}$

17) 6
18) 18
19) $\frac{6}{5}$
20) $\frac{1}{18}$
21) $\frac{2}{3}$
22) $\frac{1}{10}$

Dividing Mixed Number

1) $\frac{125}{126}$
2) $\frac{5}{2}$
3) $1\frac{27}{64}$
4) $\frac{26}{27}$
5) $2\frac{12}{49}$
6) $1\frac{19}{35}$
7) $1\frac{34}{35}$
8) $\frac{2}{81}$

9) $1\frac{2}{21}$
10) $2\frac{16}{33}$
11) $\frac{16}{29}$
12) $2\frac{1}{2}$
13) $4\frac{19}{26}$
14) $1\frac{73}{80}$
15) $3\frac{2}{3}$
16) $1\frac{1}{8}$

17) $1\frac{1}{3}$
18) $1\frac{7}{26}$
19) $3\frac{15}{28}$
20) $3\frac{1}{15}$
21) $1\frac{37}{63}$
22) $2\frac{19}{33}$
23) $\frac{5}{7}$
24) $\frac{29}{30}$

Comparing Fractions

1) >
2) >
3) <
4) >
5) <
6) <

7) >
8) <
9) <
10) >
11) >
12) >

13) <
14) <
15) >
16) <
17) >
18) >

19) >
20) <
21) >
22) >
23) =
24) >

Chapter 3:

Decimal

Round Decimals

Round each number to the correct place value

1) 0.<u>7</u>3 =

2) 5.0<u>2</u> =

3) 10.<u>7</u>11 =

4) 0.<u>4</u>67 =

5) <u>8</u>.924 =

6) 0.0<u>7</u>5 =

7) 8.<u>1</u>2 =

8) 63.7<u>4</u>0 =

9) 2.5<u>3</u>8 =

10) 12.<u>2</u>97 =

11) 2.<u>0</u>8 =

12) 5.<u>3</u>24 =

13) 2.<u>1</u>32 =

14) 8.0<u>7</u>32 =

15) 7<u>5</u>.78 =

16) 4<u>8</u>.24 =

17) 6<u>2</u>7.132 =

18) 624.<u>7</u>88 =

19) 17.4<u>8</u>1 =

20) 9<u>4</u>.86 =

21) 4.3<u>0</u>67 =

22) 57.<u>0</u>86 =

23) 224.<u>2</u>24 =

24) 0.1<u>3</u>44 =

25) 0.00<u>7</u>8 =

26) 7.0<u>3</u>67 =

27) 15.4<u>4</u>33 =

28) 21.0<u>9</u>31 =

Decimals Addition

Add the following.

1) $\begin{array}{r} 34.21 \\ + 14.25 \\ \hline \end{array}$

2) $\begin{array}{r} 0.66 \\ + 0.31 \\ \hline \end{array}$

3) $\begin{array}{r} 25.36 \\ + 20.87 \\ \hline \end{array}$

4) $\begin{array}{r} 75.165 \\ + 4.105 \\ \hline \end{array}$

5) $\begin{array}{r} 8.650 \\ + 7.82 \\ \hline \end{array}$

6) $\begin{array}{r} 5.324 \\ + 2.138 \\ \hline \end{array}$

7) $\begin{array}{r} 81.21 \\ + 15.85 \\ \hline \end{array}$

8) $\begin{array}{r} 71.05 \\ + 11.35 \\ \hline \end{array}$

9) $\begin{array}{r} 26.21 \\ + 8.07 \\ \hline \end{array}$

10) $\begin{array}{r} 6.96 \\ + 13.23 \\ \hline \end{array}$

11) $\begin{array}{r} 15.214 \\ + 11.251 \\ \hline \end{array}$

12) $\begin{array}{r} 72.36 \\ + 5.32 \\ \hline \end{array}$

13) $\begin{array}{r} 52.05 \\ + 10.54 \\ \hline \end{array}$

14) $\begin{array}{r} 107.11 \\ + 5.05 \\ \hline \end{array}$

Decimals Subtraction

Subtract the following

1) $\begin{array}{r} 7.45 \\ -\ 5.12 \\ \hline \end{array}$

8) $\begin{array}{r} 42.56 \\ -\ 22.45 \\ \hline \end{array}$

2) $\begin{array}{r} 85.35 \\ -\ 72.37 \\ \hline \end{array}$

9) $\begin{array}{r} 58.13 \\ -\ 32.35 \\ \hline \end{array}$

3) $\begin{array}{r} 0.82 \\ -\ 0.6 \\ \hline \end{array}$

10) $\begin{array}{r} 8.763 \\ -\ 0.425 \\ \hline \end{array}$

4) $\begin{array}{r} 11.245 \\ -\ 8.6 \\ \hline \end{array}$

11) $\begin{array}{r} 55.69 \\ -\ 45.32 \\ \hline \end{array}$

5) $\begin{array}{r} 0.652 \\ -\ 0.09 \\ \hline \end{array}$

12) $\begin{array}{r} 10.352 \\ -\ 4.325 \\ \hline \end{array}$

6) $\begin{array}{r} 75.25 \\ -\ 28.88 \\ \hline \end{array}$

13) $\begin{array}{r} 11.105 \\ -\ 3.128 \\ \hline \end{array}$

7) $\begin{array}{r} 112.66 \\ -\ 88.98 \\ \hline \end{array}$

14) $\begin{array}{r} 126.78 \\ -\ 8.52 \\ \hline \end{array}$

Decimals Multiplication

Solve.

1) $\begin{array}{r} 2.1 \\ \times\, 4.4 \\ \hline \end{array}$

2) $\begin{array}{r} 5.2 \\ \times\; 3.7 \\ \hline \end{array}$

3) $\begin{array}{r} 7.04 \\ \times\, 3.04 \\ \hline \end{array}$

4) $\begin{array}{r} 55.02 \\ \times\; 100 \\ \hline \end{array}$

5) $\begin{array}{r} 61.8 \\ \times\; 10 \\ \hline \end{array}$

6) $\begin{array}{r} 35.62 \\ \times\; 5.5 \\ \hline \end{array}$

7) $\begin{array}{r} 32.75 \\ \times\, 11.3 \\ \hline \end{array}$

8) $\begin{array}{r} 1.65 \\ \times\, 7.35 \\ \hline \end{array}$

9) $\begin{array}{r} 10.05 \\ \times\, 0.06 \\ \hline \end{array}$

10) $\begin{array}{r} 21.04 \\ \times\; 6.08 \\ \hline \end{array}$

11) $\begin{array}{r} 10.34 \\ \times\; 11.2 \\ \hline \end{array}$

12) $\begin{array}{r} 7.67 \\ \times\, 0.05 \\ \hline \end{array}$

13) $\begin{array}{r} 7.2 \\ \times\, 0.16 \\ \hline \end{array}$

14) $\begin{array}{r} 13.2 \\ \times\, 4.05 \\ \hline \end{array}$

Decimal Division

Dividing Decimals.

1) $7 \div 10{,}000 =$

2) $6 \div 100 =$

3) $7.1 \div 100 =$

4) $0.004 \div 10 =$

5) $9 \div 81 =$

6) $8 \div 64 =$

7) $5 \div 45 =$

8) $9 \div 180 =$

9) $7 \div 1{,}000 =$

10) $0.6 \div 0.63 =$

11) $0.9 \div 0.009 =$

12) $0.6 \div 0.12 =$

13) $0.6 \div 0.42 =$

14) $0.4 \div 0.04 =$

15) $3.08 \div 10 =$

16) $9.4 \div 10 =$

17) $6.75 \div 100 =$

18) $18.3 \div 3.3 =$

19) $64.4 \div 4 =$

20) $0.4 \div 0.004 =$

21) $7.05 \div 3.5 =$

22) $0.08 \div 0.40 =$

23) $1.8 \div 15.2 =$

24) $0.18 \div 108 =$

25) $15.72 \div 1.5 =$

26) $0.05 \div 250 =$

Comparing Decimals

Write the Correct Comparison Symbol (>, < or =)

1) 2.62 _____ 3.62

2) 0.8 _____ 0.726

3) 15.6 _____ 15.600

4) 8.07 _____ 8.70

5) 0.922 _____ 0.92

6) 0.856 _____ 0.956

7) 4.34 _____ 4.242

8) 5.0025 _____ 5.025

9) 24.087 _____ 24.078

10) 7.12 _____ 7.29

11) 4.44 _____ 4.444

12) 0.09 _____ 0.18

13) 1.302 _____ 1.32

14) 9.56 _____ 9.0569

15) 0.55 _____ 0.055

16) 61.04 _____ 61.040

17) 0.350 _____ 0.45

18) 53.92 _____ 55.01

19) 0.075 _____ 0.705

20) 46.5 _____ 39.8

21) 7.89 _____ 10.2

22) 0.014 _____ 0.0104

23) 9.042 _____ 0.9042

24) 6.5 _____ 0.658

25) 8.5 _____ 0.859

26) 6.32 _____ 6.3200

27) 1.43 _____ 0.143

28) 7.0809 _____ 7.0890

Convert Fraction to Decimal

Write each as a decimal.

1) $\dfrac{25}{50} =$

2) $\dfrac{92}{200} =$

3) $\dfrac{24}{150} =$

4) $\dfrac{32}{64} =$

5) $\dfrac{8}{72} =$

6) $\dfrac{56}{100} =$

7) $\dfrac{4}{50} =$

8) $\dfrac{31}{48} =$

9) $\dfrac{27}{300} =$

10) $\dfrac{15}{55} =$

11) $\dfrac{16}{32} =$

12) $\dfrac{12}{32} =$

13) $\dfrac{6}{20} =$

14) $\dfrac{18}{250} =$

15) $\dfrac{24}{80} =$

16) $\dfrac{30}{40} =$

17) $\dfrac{68}{100} =$

18) $\dfrac{7}{35} =$

19) $\dfrac{87}{100} =$

20) $\dfrac{1}{120} =$

21) $\dfrac{30}{180} =$

22) $\dfrac{6}{240} =$

Convert Decimal to Percent

Write each as a percent.

1) 0.285 =

2) 0.14 =

3) 3.2 =

4) 0.019 =

5) 0.007 =

6) 0.786 =

7) 0.245 =

8) 0.57 =

9) 0.002 =

10) 0.205 =

11) 0.324 =

12) 84.9 =

13) 3.015 =

14) 0.9 =

15) 7.35 =

16) 0.0312 =

17) 0.0061 =

18) 0.960 =

19) 6.68 =

20) 0.484 =

21) 8.957 =

22) 0.879 =

23) 2.7 =

24) 0.7 =

25) 2.6 =

26) 36.2 =

27) 1.52 =

28) 0.008 =

Convert Fraction to Percent

Write each as a percent.

1) $\dfrac{2}{8} =$

2) $\dfrac{3}{8} =$

3) $\dfrac{14}{28} =$

4) $\dfrac{30}{70} =$

5) $\dfrac{12}{28} =$

6) $\dfrac{17}{68} =$

7) $\dfrac{8}{11} =$

8) $\dfrac{14}{30} =$

9) $\dfrac{6}{50} =$

10) $\dfrac{12}{48} =$

11) $\dfrac{5}{34} =$

12) $\dfrac{81}{30} =$

13) $\dfrac{48}{160} =$

14) $\dfrac{32}{50} =$

15) $\dfrac{16}{58} =$

16) $\dfrac{2}{22} =$

17) $\dfrac{32}{88} =$

18) $\dfrac{21}{36} =$

19) $\dfrac{18}{92} =$

20) $\dfrac{8}{80} =$

21) $\dfrac{72}{900} =$

22) $\dfrac{360}{180} =$

Answers of Worksheets

Round Decimals

1) 0.7	11) 2.1	21) 4.31
2) 5.0	12) 5.3	22) 57.1
3) 10.7	13) 2.1	23) 224.2
4) 0.5	14) 8.07	24) 0.13
5) 9.0	15) 76.0	25) 0.008
6) 0.08	16) 48.0	26) 7.04
7) 8.1	17) 630.0	27) 15.44
8) 63.74	18) 624.8	28) 21.09
9) 2.54	19) 17.48	
10) 12.3	20) 95.0	

Decimals Addition

1) 48.46	6) 7.462	11) 26.465
2) 0.97	7) 97.06	12) 77.68
3) 46.23	8) 82.4	13) 62.59
4) 79.27	9) 34.28	14) 112.16
5) 16.47	10) 20.19	

Decimals Subtraction

1) 2.33	6) 46.37	11) 10.37
2) 12.98	7) 23.68	12) 6.027
3) 0.22	8) 20.11	13) 7.977
4) 2.645	9) 25.78	14) 118.26
5) 0.562	10) 8.338	

Decimals Multiplication

1) 9.24	6) 195.91	11) 115.808
2) 19.24	7) 370.075	12) 0.3835
3) 21.4016	8) 12.1275	13) 1.152
4) 5,502	9) 0.603	14) 53.46
5) 618	10) 127.9232	

Decimal Division

1) 0.0007	2) 0.06	3) 0.071

4) 0.0004

5) 0.111….

6) 0.125

7) 0.111…

8) 0.05

9) 0.007

10) 0.952…

11) 100

12) 5

13) 1.4285…

14) 10

15) 0.308

16) 0.94

17) 0.0675

18) 5.5454…

19) 16.1

20) 100

21) 2.01428…

22) 0.2

23) 0.1184…

24) 0.0017

25) 10.48

26) 0.0002

Comparing Decimals

1) <

2) >

3) =

4) <

5) >

6) <

7) >

8) <

9) >

10) >

11) <

12) <

13) <

14) >

15) >

16) =

17) <

18) <

19) <

20) >

21) <

22) >

23) >

24) >

25) >

26) =

27) >

28) <

Convert Fraction to Decimal

1) 0.5

2) 0.46

3) 0.16

4) 0.5

5) 0.11

6) 0.56

7) 0.08

8) 0.646

9) 0.09

10) 0.27

11) 0.5

12) 0.375

13) 0.3

14) 0.072

15) 0.3

16) 0.75

17) 0.68

18) 0.2

19) 0.87

20) 0.0083

21) 0.166

22) 0.025

Convert Decimal to Percent

1) 28.5%

2) 14%

3) 320%

4) 1.9%

5) 0.7%

6) 78.6%

7) 24.5%

8) 57%

9) 0.2%

10) 20.5%

11) 32.4%

12) 8,490%

13) 301.5%

14) 90%

15) 735%

16) 3.12%

17) 0.61%

18) 96%

19) 668%

20) 48.4%

21) 895.7%

22) 87.9%

23) 270%

24) 70%

25) 260%

26) 3,620%

27) 152%

28) 0.8%

Convert Fraction to Percent

1) 25%

2) 37.5%

3) 50%

4) 42.86%

5) 29.31%

6) 25%

7) 72.72%

8) 46.66%

9) 12%

10) 25%

11) 14.7%

12) 2.7%

13) 30%

14) 64%

15) 27.58%

16) 9.09%

17) 36.36%

18) 58.33%

19) 19.56%

20) 10%

21) 8%

22) 200%

Chapter 4:
Equations and
Inequality

Distributive and Simplifying Expressions

Simplify each expression.

1) $4x + 4 - 9 =$

2) $-(-5 - 7x) =$

3) $(-2x + 5)(-3) =$

4) $(-3x)(x + 4) =$

5) $-2x + x^2 + 4x^2 =$

6) $7y + 7x + 8y - 5x =$

7) $-3x + 3y + 14x - 9y =$

8) $-2x - 5 + 8x + \frac{16}{4} =$

9) $5 - 8(x - 2) =$

10) $-5 - 5x + 3x =$

11) $(x - 3y)2 + 4y =$

12) $2.5x^2 \times (-5x) =$

13) $-4 - 2x^2 + 6x^2 =$

14) $8 + 14x^2 + 4 =$

15) $2(-2x - 5) + 12 =$

16) $(-x)(-2 + 4x) - x(6 + x) =$

17) $-3(6 + 12) - 3x + 5x =$

18) $-4(5 - 12x - 3x) =$

19) $3(-2x - 6) =$

20) $21 + 8x - 21 =$

21) $x(-4x + 7) =$

22) $5xy + 4x - 3y + x + 2y =$

23) $3(-x - 7) + 9 =$

24) $(-3x - 8) + 12 =$

25) $3x + 4y - 6 + 1 =$

26) $(-2 + 4x) - 4x(1 + 3x) =$

27) $(-4)(-2x - 2y) =$

28) $6(-x - 3) + 4 =$

Factoring Expressions

Factor the common factor out of each expression.

1) $15x - 12 =$

2) $3x - 12 =$

3) $\frac{45}{15}x - 24 =$

4) $6b - 30 =$

5) $4a^2 - 24a =$

6) $2xy - 10y =$

7) $5x^2y + 15x =$

8) $a^2 - 8a + 7ab =$

9) $2a^2 + 2ab =$

10) $4x + 20 =$

11) $24x - 36xy =$

12) $8x - 6 =$

13) $\frac{1}{4}x - \frac{3}{4}y =$

14) $7xy - \frac{14}{3}x =$

15) $4ab + 12c =$

16) $\frac{1}{5}x - \frac{4}{5} =$

17) $12x - 18xy =$

18) $x^2 + 8x =$

19) $4x^2 - 12y =$

20) $4x^3 + 3xy + x^2 =$

21) $21x - 14 =$

22) $20b - 60c + 20d =$

23) $24ab - 8ac =$

24) $ax - ay - 3x + 3y =$

25) $3ax + 4a + 9x + 12 =$

26) $x^2 - 15x =$

27) $7x^3 - 14x^2 =$

28) $5x^2 - 60xy =$

Evaluate One Variable Expressions

Evaluate each using the values given.

1) $x + 5x, x = 2$

2) $5(-7 + 4x), x = 1$

3) $4x + 6x, x = -1$

4) $4(2 - x) + 4, x = 2$

5) $8x + 2x - 12, x = 2$

6) $5x + 11x + 12, x = -1$

7) $5x - 2x - 4, x = 5$

8) $\frac{3(5x+8)}{9}, x = 2$

9) $2x - 85, x = 32$

10) $\frac{x}{18}, \; x = 108$

11) $7(3 + 2x) - 33, x = 5$

12) $7(x + 3) - 23, x = 4$

13) $\frac{x+(-6)}{-3}, x = -6$

14) $8(6 - 3x) + 5, x = 2$

15) $-5 - \frac{2x}{10} + 6x, x = 10$

16) $5x + 11x, x = 1$

17) $-12x + 3(5 + 3x), x = -7$

18) $x + 11x, x = 0.5$

19) $\frac{(2x-2)}{6}, x = 13$

20) $3(-1 - 2x), x = 5$

21) $5x - (5 - x), x = 3$

22) $\left(-\frac{15}{x}\right) + 2 + x, x = 5$

23) $-\frac{x \times 7}{x}, x = 8$

24) $2(-1 - 3x), x = 2$

25) $3x^2 + 8x, x = 1$

26) $2(5x + 2) - 3(x - 3), x = 2$

27) $-5x - 5, x = -5$

28) $6x + 3x, x = 2$

Evaluate Two Variable Expressions

Evaluate the expressions.

1) $x + 6y$, $x = 3, y = 5$

2) $(-2)(-2x - 3y)$, $x = 1, y = 1$

3) $3x + 4y$, $x = 6, y = 2$

4) $\frac{x-4}{y+1}$, $x = 18, y = 1$

5) $\frac{a}{8} - 6b$, $a = 32, b = 4$

6) $3x - 4(y - 8)$, $x = 5, y = 3$

7) $3x + 2y - 10$, $x = 2, y = 10$

8) $-3x + 10 + 8y - 5$, $x = 2, y = 1$

9) $yx \div 3$, $x = 9, y = 9$

10) $a - b \div 3$, $a = 3, b = 12$

11) $6(x - y)$, $x = 7, y = 4$

12) $5x - 4y$, $x = 5, y = 8$

13) $\frac{10}{a} + 3b$, $a = 5, b = 4$

14) $2x^2 + 4xy$, $x = 3, y = 5$

15) $10 - \frac{xy}{12} + y$, $x = 4, y = 3$

16) $5(3x - y)$, $x = 4, y = -6$

17) $5x^2 - 3y^2$, $x = -1, y = 2$

18) $5x + \frac{y}{4}$, $x = 6, y = 16$

19) $4(4x - 2y)$, $x = 3, y = 5$

20) $4x(y - \frac{1}{2})$, $x = 5, y = 4$

21) $5(x^2 - 2y)$, $x = 3, y = 2$

22) $5xy$, $x = 2, y = 8$

23) $\frac{1}{3}y^3(y - \frac{1}{4}x)$, $x = -4, y = 3$

24) $-3(x - 5y) - 2x$, $x = 4, y = 2$

25) $-2x + \frac{1}{6}xy$, $x = 3, y = 6$

26) $x^2 + xy^2$, $x = 6, y = 5$

27) $x - 3y + 6$, $x = 8, y = 5$

28) $\frac{xy}{4x+y}$, $x = 6, y = 5$

Graphing Linear Equation

Sketch the graph of each line.

1) $y = 2x - 5$ 2) $y = -2x + 3$ 3) $x - y = 0$

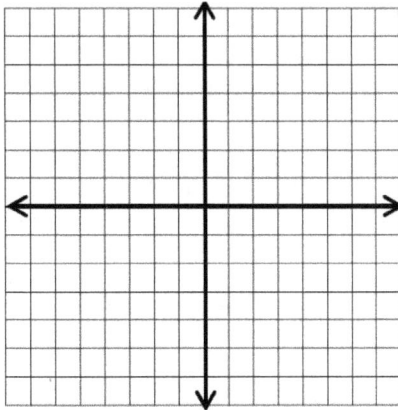 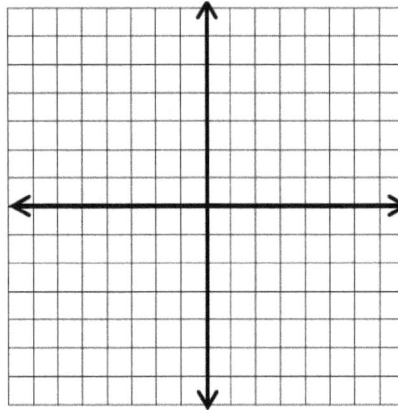 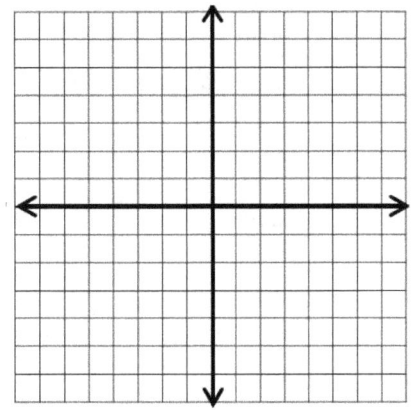

4) $x + y = 3$ 5) $5x + 3y = -2$ 6) $y - 3x + 2 = 0$

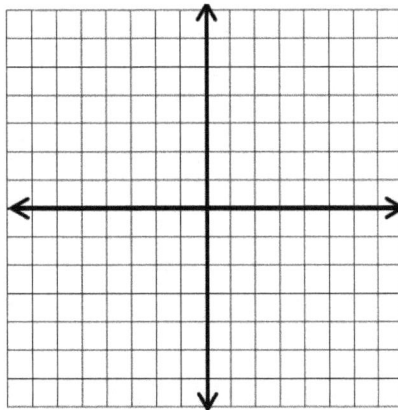 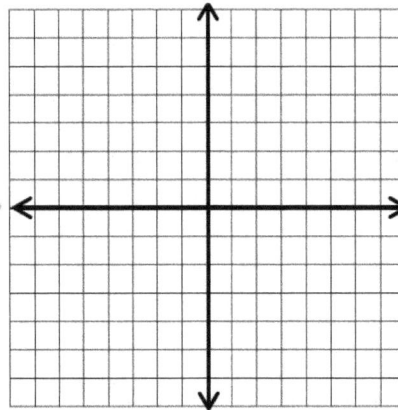 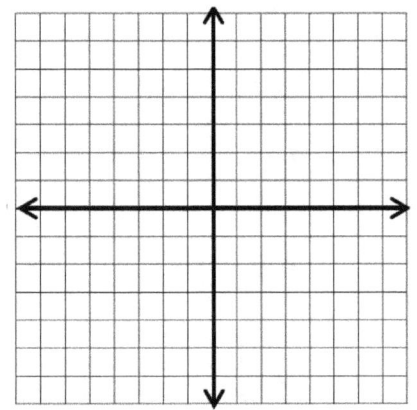

One Step Equations

Solve each equation.

1) $88 = (-24) + x$

2) $6x = (-81)$

3) $(-63) = (-7x)$

4) $(-8) = 3 + x$

5) $4 + \frac{x}{2} = (-3)$

6) $8x = (-104)$

7) $62 = x - 13$

8) $\frac{x}{3} = (-15)$

9) $x + 112 = 154$

10) $x - \frac{1}{3} = \frac{2}{3}$

11) $(-24) = x - 32$

12) $(-3x) = 39$

13) $(-169) = (13x)$

14) $-4x + 42 = 50$

15) $5x + 3 = 38$

16) $80 = (-8x)$

17) $3x + 7 = 19$

18) $11x = 121$

19) $x - 18 = 15$

20) $0.9x = 4.5$

21) $4x = 84$

22) $2x + 2.98 = 66.98$

23) $x + 9 = 6$

24) $x + 24 = 16$

25) $9x + 51 = 15$

26) $\frac{1}{6}x + 60 = 48$

Two Steps Equations

Solve each equation.

1) $12(3 + x) = 84$

2) $(-14)(x - 2) = 112$

3) $(-4)(3x - 4) = (-8)$

4) $15(2 + x) = -45$

5) $38(3x + 11) = 76$

6) $4(2x + 2) = 24$

7) $5(8 + 3x) = (-20)$

8) $(-5)(5x - 3) = 40$

9) $2x + 12 = 16$

10) $\frac{4x - 5}{5} = 3$

11) $(-3) = \frac{x + 4}{7}$

12) $80 = (-8)(x - 3)$

13) $\frac{x}{3} + 27 = 39$

14) $\frac{1}{8} = \frac{1}{4} + \frac{x}{8}$

15) $\frac{33 + 3x}{15} = (-6)$

16) $(-3)(10 + 5x) = (-15)$

17) $(-3x) + 12 = 24$

18) $\frac{x + 5}{5} = -5$

19) $\frac{x + 23}{8} = 3$

20) $(-4) + \frac{x}{2} = (-14)$

21) $-5 = \frac{x + 8}{6}$

22) $\frac{27x - 9}{18} = 4$

23) $\frac{2x - 12}{4} = 3$

24) $45 = (-5)(x - 45)$

Multi Steps Equations

Solve each equation.

1) $5 - (4 - 5x) = 6$

2) $-25 = -(4x + 17)$

3) $6x - 22 = (-2x) + 10$

4) $-75 = (-5x) - 10x$

5) $3(2 + 3x) + 3x = -30$

6) $5x - 18 = 2 + 2x - 7 + 2x$

7) $12 - 6x = (-36) - 3x + 3x$

8) $16 - 4x - 4x = 8 - 4x$

9) $8 + 7x + x = (-12) + 3x$

10) $(-3x) - 3(-2 + 4x) = 366$

11) $20 = (-200x) - 5 + 5$

12) $61 = 5x - 23 + 7x$

13) $14(4 + 2x) = 280$

14) $-60 = (-7x) - 13x$

15) $2(4x + 5) = -2(x + 4) - 22$

16) $11x - 17 = 6x + 8$

17) $9 = -3(x - 8)$

18) $(-6) - 8x = 6(1 + 2x)$

19) $x + 3 = -2(9 + 3x)$

20) $15 = 6 - 5x - 11$

21) $-15 - 9x - 3x = 12 - 3x$

22) $-30 - 3x + 5x = 20 - 23x$

23) $35 - 6x - 16x = -10 - 16x$

24) $15x - 17 = 6x + 10$

Graphing Linear Inequalities

Sketch the graph of each linear inequality.

1) $y > 2x - 3$　　　　2) $y < x + 3$　　　　3) $y \leq -3x - 8$

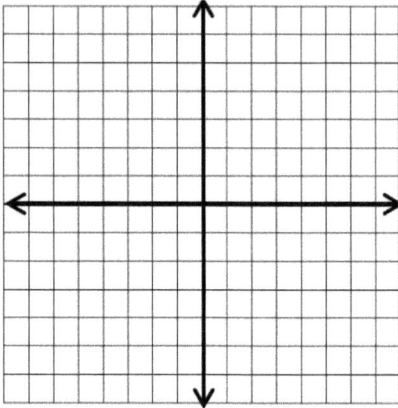　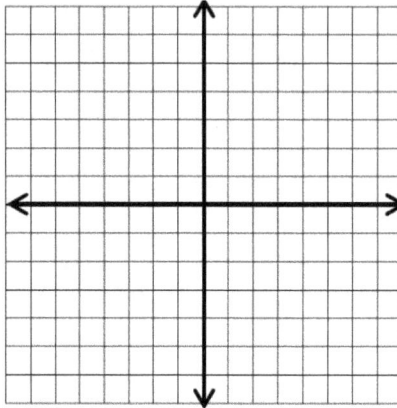　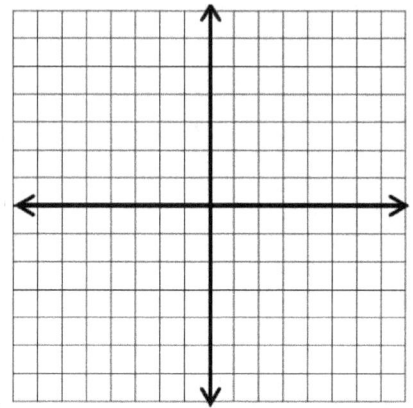

4) $3y \geq 6 + 3x$　　　　5) $-3y < x - 12$　　　　6) $2y \geq -8x + 4$

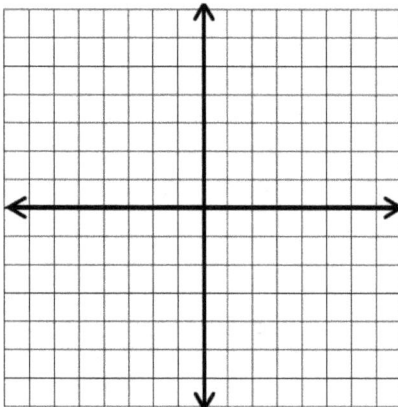　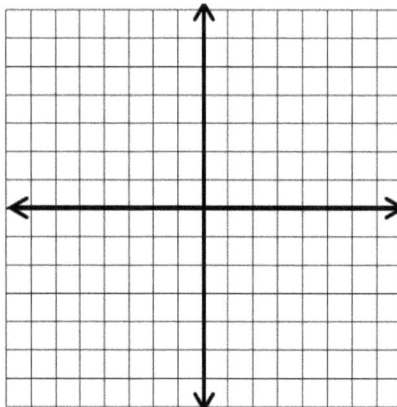　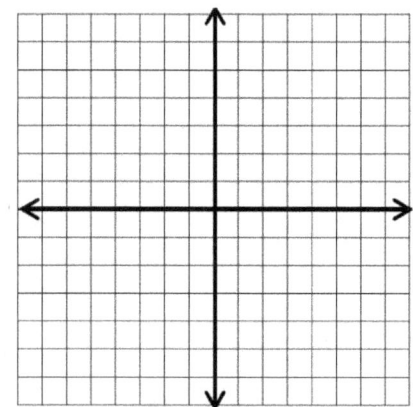

One Step Inequality

Solve each inequality.

1) $14x < 28$

2) $x + 17 \geq -4$

3) $x - 2 \leq 10$

4) $-2x + 4 > -10$

5) $x + 18 \geq -6$

6) $x + 9 \geq 5$

7) $x - \frac{1}{3} \leq 5$

8) $-7x < 42$

9) $-x + 8 > -3$

10) $\frac{x}{3} + 3 > -9$

11) $-x + 8 > -4$

12) $x - 14 \leq 18$

13) $-x - 7 \leq -8$

14) $x + 28 \geq -11$

15) $x + \frac{1}{3} \geq -\frac{2}{3}$

16) $x + 6 \geq -14$

17) $x - 42 \leq -48$

18) $x - 5 \leq 4$

19) $-x + 5 > -6$

20) $x + 6 \geq -12$

21) $8x + 6 \leq 22$

22) $6x - 3 \geq 9$

23) $9x - 5 < 22$

24) $6x - 7 \leq 35$

Two Steps Inequality

Solve each inequality

1) $4x - 6 \leq 14$

2) $9x - 12 \leq 24$

3) $\frac{-1}{4}x + \frac{x}{2} \leq \frac{1}{8}$

4) $10x + 20 \geq 60$

5) $8x - 14 \geq 18$

6) $3x - 5 \leq 16$

7) $8x - 2 \leq 14$

8) $9x + 5 \leq 23$

9) $2x + 10 > 32$

10) $\frac{x}{8} + 2 \leq 4$

11) $3x + 4 \geq 37$

12) $3x - 8 < 10$

13) $6 \geq \frac{x+7}{2}$

14) $3x + 9 < 48$

15) $\frac{4+x}{5} \geq 3$

16) $16 + 4x < 36$

17) $16 > 6x - 8$

18) $11 + \frac{x}{2} < 7$

19) $-8 + 8x > 48$

20) $6 + \frac{x}{9} < 3$

Multi Steps Inequality

Solve the inequalities.

1) $8x - 12 < 10x - 18$

2) $\frac{4x + 10}{6} \le x$

3) $14x - 10 > 6x + 30$

4) $-3x > -6x + 4$

5) $3 + \frac{x}{2} < \frac{x}{4}$

6) $\frac{4x - 6}{8} > x$

7) $4x - 20 + 4 > 6x - 8$

8) $x - 8 > 11 + 3(x + 5)$

9) $\frac{x}{3} + 2 > x$

10) $-7x + 8 \ge -6(4x - 8) - 8x$

11) $7x - 4 \le 8x + 9$

12) $\frac{2x - 7}{5} > 2$

13) $8(x + 2) < 6x + 10$

14) $-8x + 12 \le 4(x - 9)$

15) $\frac{5x - 6}{3} > 3x + 2$

16) $2(x - 8) + 10 \ge 4x - 2$

17) $\frac{-5x+7}{6} > 5x$

18) $-6x - 8 > -14x$

19) $\frac{1}{4}x - 16 > \frac{1}{8}x - 23$

20) $-16(x - 9) \le 20x$

Systems of Equations

Calculate each system of equations.

1) $-12x + 14y = 16$ $x =$ ____
    $2x + 8y = 18$ $y =$ ____

2) $-8x + 24y = 24$ $x =$ ____
    $28x - 32y = 20$ $y =$ ____

3) $y = -9$ $x =$ ____
    $4x - 10y = 24$ $y =$ ____

4) $8y = -8x + 40$ $x =$ ____
    16
    $x - 4y = -24$ $y =$ ____

5) $10x - 9y = -13$ $x =$ ____
    $-5x + 3y = 11$ $y =$ ____

6) $-6x - 8y = 10$ $x =$ ____
    $4x - 8y = 20$ $y =$ ____

7) $5x - 14y = -23$ $x =$ ____
    $-6x + 7y = 8$ $y =$ ____

8) $-4x + 3y = 3$ $x =$ ____
    $-x + 2y = 5$ $y =$ ____

9) $-4x + 5y = 15$ $x =$ ____
    $-3x + 4y = -10$ $y =$ ____

10) $-6x - 6y = -21$ $x =$ ____
    $-6x + 6y = -66$ $y =$ ____

11) $12x - 21y = 6$ $x =$ ____
    $-6x - 3y = -12$ $y =$ ____

12) $-8x - 8y = -28$ $x =$ ____
    $8x - 8y = 88$ $y =$ ____

13) $8x + 10y = 6$ $x =$ ____
    $6x - 2y = 12$ $y =$ ____

14) $6x - 4y = 4$ $x =$ ____
    $20x - 20y = 40$ $y =$ ____

15) $10x + 16y = 28$ $x =$ ____
    $-6x - 4y = -6$ $y =$ ____

16) $16x + 10y = 8$ $x =$ ____
    $-6x - 8y = 30$ $y =$ ____

Systems of Equations Word Problems

Find the answer for each word problem.

1) Tickets to a movie cost $6 for adults and $4 for students. A group of friends purchased 9 tickets for $50.00. How many adults ticket did they buy? ____

2) At a store, Eva bought two shirts and five hats for $77.00. Nicole bought three same shirts and four same hats for $84.00. What is the price of each shirt? _____

3) A farmhouse shelters 10 animals, some are pigs, and some are ducks. Altogether there are 36 legs. How many pigs are there? _____

4) A class of 85 students went on a field trip. They took 24 vehicles, some cars and some buses. If each car holds 3 students and each bus hold 16 students, how many buses did they take? _____

5) A theater is selling tickets for a performance. Mr. Smith purchased 8 senior tickets and 10 child tickets for $248 for his friends and family. Mr. Jackson purchased 4 senior tickets and 6 child tickets for $132. What is the price of a senior ticket? $_____

6) The difference of two numbers is 15. Their sum is 33. What is the bigger number? $_____

7) The sum of the digits of a certain two–digit number is 7. Reversing its digits increase the number by 9. What is the number? _____

8) The difference of two numbers is 11. Their sum is 25. What are the numbers? _____

9) The length of a rectangle is 5 meters greater than 2 times the width. The perimeter of rectangle is 28 meters. What is the length of the rectangle? _____

10) Jim has 23 nickels and dimes totaling $2.40. How many nickels does he have? _____

Finding Distance of Two Points

Find the distance between each pair of points.

1) $(4, 2), (-2, -6)$

2) $(-8, -4), (8, 8)$

3) $(-6, 0), (30, 48)$

4) $(-8, -2), (2, 22)$

5) $(3, -2), (-6, -14)$

6) $(-6, 0), (-2, 3)$

7) $(3, 2), (11, 17)$

8) $(-6, -10), (6, -1)$

9) $(5, 9), (-11, -3)$

10) $(3, -1), (1, -3)$

11) $(6, 0), (36, 72)$

12) $(8, 4), (3, -8)$

13) $(4, 2), (-5, -10)$

14) $(-8, 10), (4, 40)$

15) $(8, 4), (-10, -20)$

16) $(-16, -4), (32, 16)$

17) $(6, 10), (-10, -20)$

18) $(-5, 4), (7, 9)$

Find the midpoint of the line segment with the given endpoints.

1) $(-4, -4), (8, 4)$

2) $(20, 4), (-4, 4)$

3) $(6, -2), (2, 10)$

4) $(-6, -5), (2, 1)$

5) $(3, -2), (5, -2)$

6) $(-10, -4), (6, -2)$

7) $(4, 1), (-4, 9)$

8) $(-5, 6), (-5, 2)$

9) $(-8, 8), (4, -2)$

10) $(1, 7), (5, -1)$

11) $(-9, 5), (5, 3)$

12) $(7, 10), (-3, -6)$

13) $(-8, 14), (-8, 2)$

14) $(16, 7), (6, -3)$

15) $(5, 18), (-3, 12)$

16) $(-18, -1), (-10, 7)$

17) $(13, 9), (33, 27)$

18) $(-16, -22), (36, -2)$

Answers of Worksheets

Distributive and Simplifying Expressions

1) $4x - 5$
2) $5 + 7x$
3) $6x - 15$
4) $-3x^2 - 12x$
5) $5x^2 - 2x$
6) $2x + 15y$
7) $11x - 6y$
8) $6x - 1$
9) $-8x + 21$
10) $-2x - 5$

11) $2x - 2y$
12) $-12.5x^3$
13) $4x^2 - 4$
14) $14x^2 + 12$
15) $-4x + 2$
16) $-5x^2 - 4x$
17) $2x - 54$
18) $60x - 20$
19) $-6x - 18$
20) $8x$

21) $-4x^2 + 7x$
22) $5x + y + 5xy$
23) $-3x - 12$
24) $-3x + 4$
25) $3x + 4y - 5$
26) $-12x^2 - 2$
27) $8x + 8y$
28) $-6x - 14$

Factoring Expressions

1) $3(5x - 4)$
2) $3(x - 4)$
3) $3(x - 8)$
4) $6(b - 5)$
5) $4a(a - 6)$
6) $2y(x - 5)$
7) $5x(xy + 3)$
8) $a(a - 8 + 7b)$
9) $2a(a + b)$
10) $4(x + 5)$

11) $12x(2 - 3y)$
12) $2(4x - 3)$
13) $\frac{1}{4}(x - 3y)$
14) $7x(y - \frac{2}{3})$
15) $4(ab + 3c)$
16) $\frac{1}{5}(x - 4)$
17) $6x(2 - 3y)$
18) $x(x + 8)$
19) $4(x^2 - 3y)$

20) $x(4x^2 + 3y + x)$
21) $7(3x - 2)$
22) $20(b - 3c + d)$
23) $8a(3b - c)$
24) $(x - y)(a - 3)$
25) $(3x + 4)(a + 3)$
26) $x(x - 15)$
27) $7x^2(x - 2)$
28) $5x(x - 12y)$

Evaluate One Variable Expressions

1) 12
2) -15
3) -10
4) 4
5) 8
6) -4

7) 11
8) 6
9) -21
10) 6
11) 58
12) 26

13) 4
14) 5
15) 53
16) 16
17) 36
18) 6

19) 4
20) -33
21) 13
22) 4
23) -7
24) -14

25) 11	26) 27	27) 20	28) 18

Evaluate Two Variable Expressions

1) 33	9) 27	17) −7	24) 10
2) 10	10) 3	18) 34	25) −3
3) 26	11) 18	19) 8	26) 186
4) 7	12) 20	20) 70	27) −1
5) −20	13) 17	21) 25	28) $\frac{30}{29}$
6) 35	14) 78	22) 80	
7) 16	15) 12	23) 36	
8) 7	16) 90		

Graphing Lines Using Line Equation

1) $y = 2x - 5$

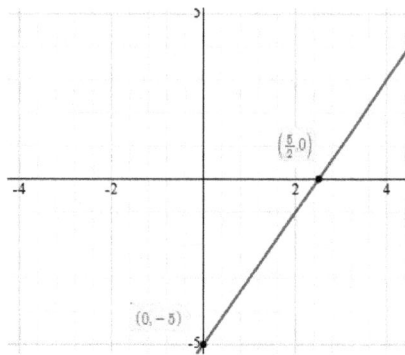

2) $y = -2x + 3$

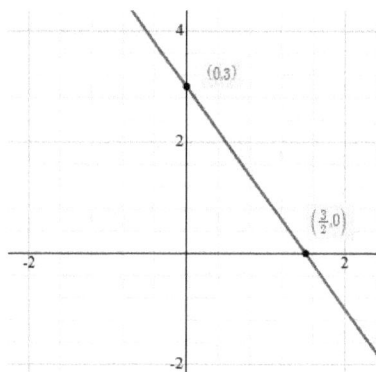

3) $x - y = 0$

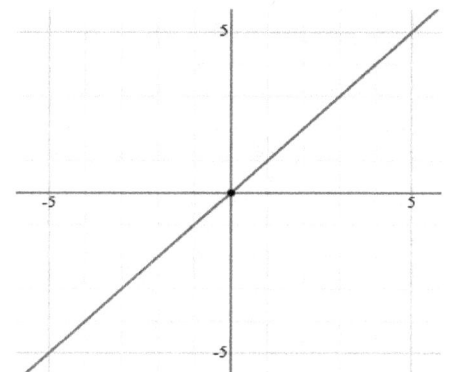

4) $x + y = 3$

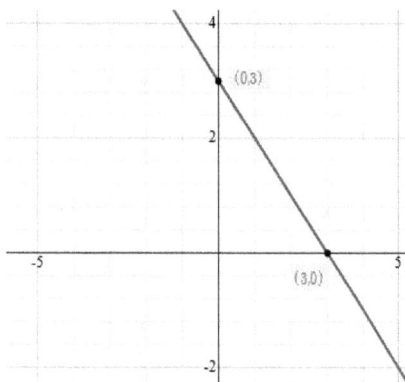

5) $5x + 3y = -2$

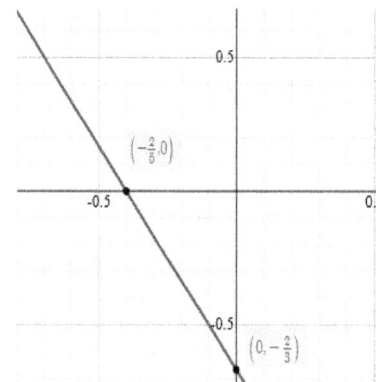

6) $y - 3x + 2 = 0$

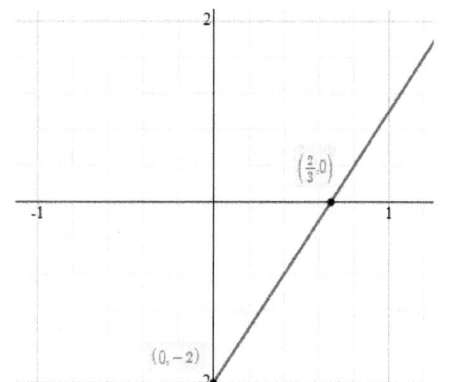

GED Math Exercise Book

One Step Equations

1) $x = 112$
2) $x = -9$
3) $x = 9$
4) $x = -11$
5) $x = -14$
6) $x = -13$
7) x = 75
8) x = -45
9) x = 42

10) x = 1
11) x = 8
12) x = -13
13) x = -13
14) x = -2
15) x = 7
16) x = -10
17) $x = 4$
18) x = 11

19) x = 33
20) x = 5
21) x = 21
22) x = 32
23) $x = -3$
24) -8
25) -4
26) -72

Two Steps Equations

1) $x = 4$
2) $x = -6$
3) $x = 2$
4) $x = -5$
5) $x = -3$
6) $x = 2$
7) $x = -4$
8) $x = -1$

9) $x = 2$
10) $x = 5$
11) $x = -25$
12) $x = -7$
13) $x = 36$
14) $x = -1$
15) $x = -41$
16) $x = -1$

17) $x = -4$
18) $x = -30$
19) $x = 1$
20) $x = -20$
21) $x = -38$
22) $x = 3$
23) $x = 12$
24) $x = 36$

Multi Steps Equations

1) $x = 1$
2) $x = 2$
3) $x = 4$
4) $x = 5$
5) $x = -3$
6) $x = 13$
7) $x = 8$
8) $x = 2$

9) $x = -4$
10) $x = -24$
11) $x = -0.1$
12) $x = 7$
13) $x = 8$
14) $x = 3$
15) $x = -4$
16) $x = 5$

17) $x = 5$
18) $x = -3/5$
19) $x = -3$
20) $x = -3$
21) $x = -4$
22) $x = 2$
23) $x = 5$
24) $x = 3$

Graphing Linear Inequalities

1) $y > 2x - 3$

2) $y < x + 3$

3) $y \leq -3x - 8$

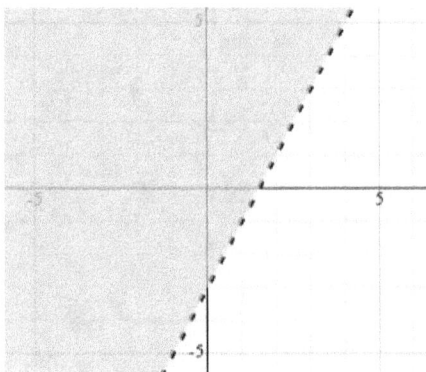
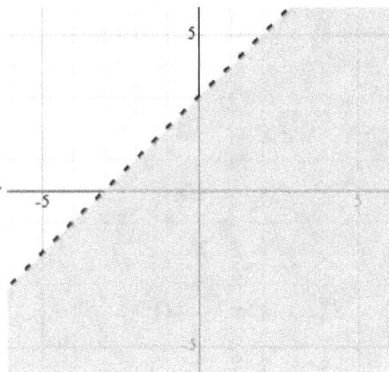
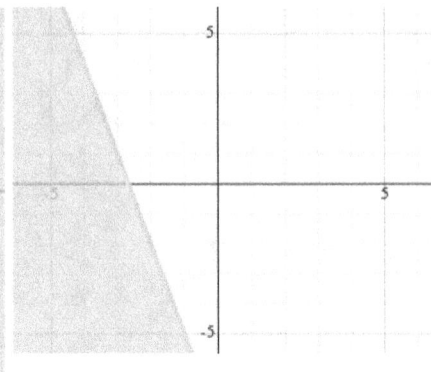

4) $3y \geq 6 + 3x$

5) $-3y < x - 12$

6) $2y \geq -8x + 4$

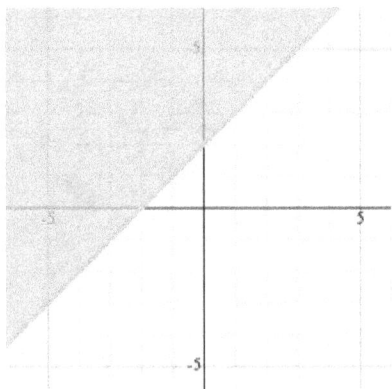
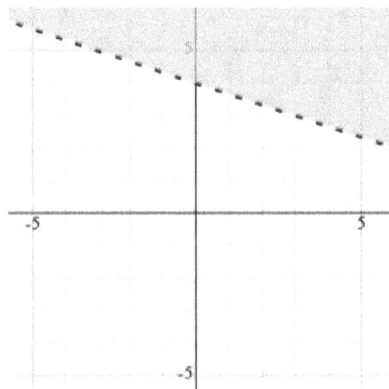
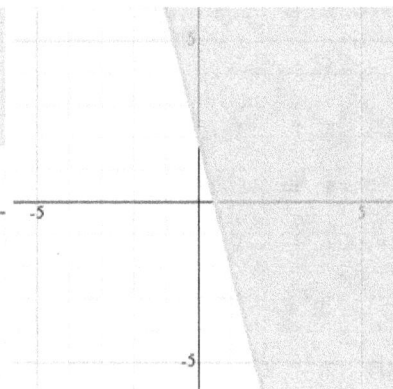

One Step Inequality

1) $x < 2$

2) $x \geq -21$

3) $x \leq 12$

4) $x < 7$

5) $x \geq -24$

6) $x \geq -4$

7) $x \leq \frac{16}{3}$

8) $x > -6$

9) $x < 11$

10) $x > -36$

11) $x < 12$

12) $x \leq 32$

13) $x \geq 1$

14) $x \geq -39$

15) $x \geq -1$

16) $x \geq -20$

17) $x \leq -6$

18) $x \leq 9$

19) $x < 11$

20) $x \geq -18$

21) $x \leq 2$

22) $x \geq 2$

23) $x < 3$

24) $x \leq 7$

Two Steps Inequality

1) $x \leq 5$

2) $x \leq 4$

3) $x \leq 0.5$

4) $x \geq 4$

5) $x \geq 4$

6) $x \leq 7$

7) $x \le 2$

8) $x \le 2$

9) $x > 11$

10) $x \le 16$

11) $x \ge 11$

12) $x < 6$

13) $x \le 5$

14) $x < 13$

15) $x \ge 11$

16) $x < 5$

17) $x < 4$

18) $x < 8$

19) $x > 7$

20) $x < -27$

Multi Steps Inequality

1) $x > 3$

2) $x \ge 5$

3) $x > 5$

4) $x > \frac{4}{3}$

5) $x < -12$

6) $x < -1.5$

7) $x < -4$

8) $x < -17$

9) $x < 3$

10) $x \ge 1.6$

11) $x \ge -13$

12) $x > 8.5$

13) $x < -3$

14) $x \ge 4$

15) $x < -3$

16) $x \le -2$

17) $x < \frac{1}{5}$

18) $x > 1$

19) $x > -56$

20) $x \ge 4$

Systems of Equations

1) $x = 1, y = 2$

2) $x = 3, y = 2$

3) $x = -\frac{33}{2}$

4) $x = -\frac{1}{5}, y = \frac{26}{5}$

5) $x = -4, y = -3$

6) $x = 1, y = -2$

7) $x = 1, y = 2$

8) $x = \frac{9}{5}, y = \frac{17}{5}$

9) $x = -110, y = -85$

10) $x = -\frac{15}{4}, y = \frac{29}{4}$

11) $x = \frac{5}{3}, y = \frac{2}{3}$

12) $x = \frac{29}{4}, y = -\frac{15}{4}$

13) $x = \frac{33}{19}, y = -\frac{15}{19}$

14) $x = -2, y = -4$

15) $x = -\frac{2}{7}, y = \frac{27}{14}$

16) $x = \frac{91}{17}, y = -\frac{132}{17}$

Systems of Equations Word Problems

1) 7

2) $16

3) 8

4) 1

5) $21

6) 24

7) 43

8) 18, 7

9) 11 meters

10) 18

Finding Distance of Two Points

1) 10

2) 20

3) 60

4) 26

5) 15

6) 5

7) 17

8) 15

9) 20

10) $2\sqrt{2}$

11) 78

12) 13

13) 15	15) 30	17) 34
14) $6\sqrt{29}$	16) 52	18) 13

Finding Midpoint

1) $(2, 0)$	7) $(0, 5)$	13) $(-8, 8)$
2) $(8, 4)$	8) $(-5, 4)$	14) $(11, 2)$
3) $(4, 2)$	9) $(-2, 3)$	15) $(1, 15)$
4) $(-2, -4)$	10) $(3, 3)$	16) $(-14, 3)$
5) $(4, -2)$	11) $(-2, 4)$	17) $(23, 18)$
6) $(-2, -3)$	12) $(2, 2)$	18) $(10, -12)$

Chapter 5:

Exponent and Radicals

Positive Exponents

Simplify. Your answer should contain only positive exponents.

1) $4^4 =$

2) $3^5 =$

3) $\frac{5x^7y}{xy} =$

4) $(11x^2x)^3 =$

5) $(x^2)^6 =$

6) $\left(\frac{1}{5}\right)^3 =$

7) $0^{10} =$

8) $6 \times 6 \times 6 =$

9) $3 \times 3 \times 3 \times 3 \times 3 =$

10) $(4x^4y)^2 =$

11) $9^3 =$

12) $(5x^3y^2)^2 =$

13) $7 \times 10^4 =$

14) $0.3 \times 0.3 \times 0.3 =$

15) $\frac{1}{4} \times \frac{1}{4} \times \frac{1}{4} =$

16) $5^5 =$

17) $(5x^8y^2)^3 =$

18) $8^3 =$

19) $y \times y \times y \times y =$

20) $8 \times 8 \times 8 \times 8 =$

21) $(2x^4y^2z)^3 =$

22) $10^0 =$

23) $(10x^4y^{-1})^2 =$

24) $(3x^2y^4)^3 =$

Negative Exponents

Simplify. Leave no negative exponents.

1) $2^{-3} =$

2) $7^{-2} =$

3) $(\frac{1}{4})^{-2} =$

4) $5^{-3} =$

5) $1^{350} =$

6) $4^{-3} =$

7) $(\frac{1}{2})^{-6} =$

8) $-8y^{-4} =$

9) $(\frac{1}{y^{-5}})^{-3} =$

10) $x^{-\frac{4}{5}} =$

11) $\frac{1}{7^{-6}} =$

12) $3^{-5} =$

13) $5^{-2} =$

14) $13^{-2} =$

15) $30^{-2} =$

16) $x^{-8} =$

17) $(x^2)^{-4} =$

18) $x^{-2} \times x^{-2} \times x^{-2} \times x^{-2} =$

19) $\frac{1}{3} \times \frac{1}{3} =$

20) $100^{-2} =$

21) $100z^{-3} =$

22) $3^{-4} =$

23) $(-\frac{1}{13})^2 =$

24) $12^0 =$

25) $(\frac{1}{x})^{-12} =$

26) $10^{-2} =$

Add and subtract Exponents

Solve each problem.

1) $3^2 + 4^3 =$

2) $x^{10} + x^{10} =$

3) $5b^3 - 4b^3 =$

4) $6 + 5^2 =$

5) $9 - 6^2 =$

6) $12 + 3^2 =$

7) $5x^2 + 8x^2 =$

8) $9^2 + 2^6 =$

9) $3^6 - 4^3 =$

10) $8^2 - 10^0 =$

11) $7^2 - 4^2 =$

12) $9^2 + 3^4 =$

13) $12^2 - 5^2 =$

14) $7^2 + 7^2 =$

15) $7^3 - 5^3 =$

16) $1^{24} + 1^{28} =$

17) $4^3 - 2^3 =$

18) $5^4 - 5^2 =$

19) $7^2 - 4^2 =$

20) $5^2 + 8^2 =$

21) $4^2 + 3^4 =$

22) $18 + 2^4 =$

23) $7x^8 + 5x^8 =$

24) $9^0 + 8^2 =$

25) $8^2 + 8^2 =$

26) $6^3 + 3^2 =$

27) $(\frac{1}{2})^2 + (\frac{1}{2})^2 =$

28) $10^2 + 3^2 =$

GED Math Exercise Book

Exponent multiplication

Simplify each of the following

1) $2^7 \times 2^4 =$

2) $5^3 \times 9^0 =$

3) $8^1 \times 3^2 =$

4) $a^{-7} \times a^{-7} =$

5) $y^{-3} \times y^{-3} \times y^{-3} =$

6) $4^5 \times 5^7 \times 4^{-4} \times 5^{-6} =$

7) $6x^4y^3 \times 4x^3y^2 =$

8) $(x^3)^5 =$

9) $(x^4y^6)^5 \times (x^4y^5)^{-5} =$

10) $8^4 \times 8^2 =$

11) $a^{4b} \times a^0 =$

12) $4^2 \times 4^2 =$

13) $a^{3m} \times a^{2n} =$

14) $2a^n \times 4b^n =$

15) $5^{-3} \times 4^{-3} =$

16) $6^{10} \times 3^{10} =$

17) $(7^6)^5 =$

18) $\left(\frac{1}{6}\right)^2 \times \left(\frac{1}{6}\right)^4 \times \left(\frac{1}{6}\right)^5 =$

19) $\left(\frac{1}{9}\right)^{52} \times 9^{52} =$

20) $(4m)^{\frac{4}{5}} \times (-2m)^{\frac{4}{5}} =$

21) $(x^4y)^{\frac{1}{4}} \times (xy^3)^{\frac{1}{4}} =$

22) $(2a^mb^n)^r =$

23) $(5x^3y^2)^3 =$

24) $(x^{\frac{1}{3}}y^2)^{\frac{-1}{3}} \times (x^4y^6)^0 =$

25) $7^8 \times 7^7 =$

26) $28^{\frac{1}{6}} \times 28^{\frac{1}{3}} =$

27) $9^4 \times 3^4 =$

28) $(x^{12})^0 =$

Exponent division

Simplify. Your answer should contain only positive exponents.

1) $\dfrac{7^6}{7} =$

2) $\dfrac{51x^4}{x} =$

3) $\dfrac{a^m}{a^{2n}} =$

4) $\dfrac{3x^{-6}}{15x^{-4}} =$

5) $\dfrac{63x^9}{7x^4} =$

6) $\dfrac{17x^7}{5x^8} =$

7) $\dfrac{36x^8}{12y^3} =$

8) $\dfrac{45xy^6}{x^4y^2} =$

9) $\dfrac{3x^9}{8x} =$

10) $\dfrac{45x^7y^9}{5x^8} =$

11) $\dfrac{12x^4}{20x^9y^{12}} =$

12) $\dfrac{8yx^7}{40yx^{10}} =$

13) $\dfrac{21x^3y^2}{3x^2y^3} =$

14) $\dfrac{x^{4.75}}{x^{0.75}} =$

15) $\dfrac{9x^4y}{18xy^3} =$

16) $\dfrac{34b^3r^8}{17a^2b^5} =$

17) $\dfrac{30x^7}{15x^9} =$

18) $\dfrac{44x^5}{11x^8} =$

19) $\dfrac{6^5}{6^3} =$

20) $\dfrac{x}{x^{15}} =$

21) $\dfrac{11^7}{11^4} =$

22) $\dfrac{5xy^5}{10y^3} =$

23) $\dfrac{11x^6y}{121xy^3} =$

24) $\dfrac{64x^5}{8y^9} =$

Scientific Notation

Write each number in scientific notation.

1) 8,500,000=

2) 700 =

3) 0.000008 =

4) 587,000 =

5) 0.00139 =

6) 0.85 =

7) 0.000093 =

8) 20,000,000 =

9) 28,000,000 =

10) 230,000,000 =

11) 0.000049 =

12) 0.00002 =

13) 0.00027 =

14) 70,000 =

15) 2,870 =

16) 190,000 =

17) 0.0223 =

18) 0.7 =

19) 0.082 =

20) 310,000 =

21) 48,000 =

22) 0.000098 =

23) 0.035 =

24) 1,778 =

25) 58,792 =

26) 24,600 =

27) 34,021 =

28) 9,100,000 =

Square Roots

Find the square root of each number.

1) $\sqrt{36} =$

2) $\sqrt{0} =$

3) $\sqrt{289} =$

4) $\sqrt{484} =$

5) $\sqrt{1,600} =$

6) $\sqrt{529} =$

7) $\sqrt{0.01} =$

8) $\sqrt{10,000} =$

9) $\sqrt{0.16} =$

10) $\sqrt{0.36} =$

11) $\sqrt{0.25} =$

12) $\sqrt{1.21} =$

13) $\sqrt{784} =$

14) $\sqrt{576} =$

15) $\sqrt{676} =$

16) $\sqrt{961} =$

17) $\sqrt{1,681} =$

18) $\sqrt{0.81} =$

19) $\sqrt{0.49} =$

20) $\sqrt{0.64} =$

21) $\sqrt{1,089} =$

22) $\sqrt{2,500} =$

23) $\sqrt{8,100} =$

24) $\sqrt{12,100} =$

25) $\sqrt{2.25} =$

26) $\sqrt{2.56} =$

27) $\sqrt{1.21} =$

28) $\sqrt{0.09} =$

Simplify Square Roots

Simplify the following.

1) $\sqrt{108} =$

2) $\sqrt{116} =$

3) $\sqrt{24} =$

4) $\sqrt{99} =$

5) $\sqrt{200} =$

6) $\sqrt{45} =$

7) $8\sqrt{50} =$

8) $3\sqrt{300} =$

9) $\sqrt{24} =$

10) $2\sqrt{18} =$

11) $4\sqrt{3} + 7\sqrt{3} =$

12) $\frac{11}{4+\sqrt{5}} =$

13) $\sqrt{48} =$

14) $\frac{4}{3-\sqrt{5}} =$

15) $\sqrt{18} \times \sqrt{2} =$

16) $\frac{\sqrt{300}}{\sqrt{3}} =$

17) $\frac{\sqrt{90}}{\sqrt{18 \times 5}} =$

18) $\sqrt{80y^6} =$

19) $6\sqrt{81a} =$

20) $\sqrt{41+8} + \sqrt{9} =$

21) $\sqrt{72} =$

22) $\sqrt{432} =$

23) $\sqrt{80} =$

24) $\sqrt{192} =$

25) $\sqrt{1,280} =$

26) $\sqrt{196} =$

Answers of Worksheets

Positive Exponents

1) 256
2) 243
3) $5x^6$
4) $1,331x^9$
5) x^{12}
6) $\frac{1}{125}$
7) 0
8) 6^3

9) 3^5
10) $16x^8y^2$
11) 729
12) $25x^6y^4$
13) 70,000
14) 0.3^3
15) $\left(\frac{1}{4}\right)^3$
16) 3,125

17) $125x^{24}y^6$
18) 512
19) y^4
20) 8^4
21) $8x^{12}y^6z^3$
22) 1
23) $\frac{100x^8}{y^2}$
24) $27x^6y^{12}$

Negative Exponents

1) $\frac{1}{8}$
2) $\frac{1}{49}$
3) 16
4) $\frac{1}{125}$
5) 1
6) $\frac{1}{64}$
7) 64
8) $\frac{-8}{y^4}$
9) y^{15}
10) $\frac{1}{x^{\frac{4}{5}}}$

11) 7^6
12) $\frac{1}{243}$
13) $\frac{1}{25}$
14) $\frac{1}{169}$
15) $\frac{1}{900}$
16) $\frac{1}{x^8}$
17) $\frac{1}{x^8}$
18) $\frac{1}{x^8}$
19) $\frac{1}{3^2}$

20) $\frac{1}{10,000}$
21) $\frac{100}{z^3}$
22) $\frac{1}{81}$
23) $\frac{1}{169}$
24) 1
25) x^{12}
26) $\frac{1}{100}$

Add and subtract Exponents

1) 73
2) $2x^{10}$
3) b^3
4) 31
5) -27

6) 21
7) $13x^2$
8) 145
9) 665
10) 63

11) 33
12) 162
13) 119
14) 98
15) 218

16) 1

17) 56

18) 600

19) 33

20) 89

21) 97

22) 34

23) $12x^8$

24) 65

25) 128

26) 225

27) $\frac{1}{2}$

28) 109

Exponent multiplication

1) 2^{11}

2) 125

3) 72

4) a^{-14}

5) y^{-9}

6) 20

7) $24x^7y^5$

8) x^{15}

9) y^5

10) 8^6

11) a^{4b}

12) 4^4

13) a^{3m+2n}

14) $8(ab)^n$

15) 20^{-3}

16) 18^{10}

17) 7^{30}

18) $(\frac{1}{6})^{11}$

19) 1

20) $(-8m^2)^{\frac{4}{5}}$

21) $x^{\frac{5}{4}}y$

22) $2^r a^{mr} b^{nr}$

23) $125x^9y^6$

24) $x^{\frac{5}{4}}y$

25) 7^{15}

26) $28^{\frac{1}{2}}$

27) $27^4 = 3^{12}$

28) 1

Exponent division

1) 7^5

2) $51x^3$

3) a^{m-2n}

4) $\frac{1}{5x^2}$

5) $9x^5$

6) $\frac{17}{5x}$

7) $\frac{3x^8}{y^3}$

8) $\frac{45y^4}{x^3}$

9) $\frac{3x^8}{8}$

10) $\frac{9y^9}{x}$

11) $\frac{3}{5x^5y^{12}}$

12) $\frac{1}{5x^3}$

13) $\frac{7x}{y}$

14) x^4

15) $\frac{x^3}{2y^2}$

16) $\frac{2r^8}{a^2b^2}$

17) $\frac{2}{x^2}$

18) $\frac{4}{x^3}$

19) 6^2

20) $\frac{1}{x^{14}}$

21) 11^3

22) $\frac{1}{2}xy^2$

23) $\frac{x^5}{11y^2}$

24) $\frac{8x^5}{y^9}$

Scientific Notation

1) 8.5×10^6

2) 7×10^2

3) 8×10^{-6}

4) 5.87×10^5

5) 1.39×10^{-3}

6) 8.5×10^{-1}

7) 9.3×10^{-5}

8) 2×10^7

9) 2.8×10^7

10) 2.3×10^8

11) 4.9×10^{-5}

12) 2×10^{-5}

13) 2.7×10^{-4}

14) 7×10^4

15) 2.87×10^3

16) 1.9×10^5

17) 2.23×10^{-2}

18) 7×10^{-1}

19) 8.2×10^{-2}

20) 3.1×10^5

21) 4.8×10^4

22) 9.8×10^{-5}

23) 3.5×10^{-2}

24) 1.778×10^3

25) 5.8792×10^4

26) 2.46×10^4

27) 3.4021×10^4

28) 9.1×10^6

Square Roots

1) 6

2) 0

3) 17

4) 22

5) 40

6) 23

7) 0.1

8) 100

9) 0.4

10) 0.6

11) 0.5

12) 1.1

13) 28

14) 24

15) 26

16) 31

17) 41

18) 0.9

19) 0.7

20) 0.8

21) 33

22) 50

23) 90

24) 110

25) 1.5

26) 1.6

27) 1.1

28) 0.3

Simplify Square Roots

1) $6\sqrt{3}$

2) $2\sqrt{29}$

3) $2\sqrt{6}$

4) $3\sqrt{11}$

5) $10\sqrt{2}$

6) $3\sqrt{5}$

7) $40\sqrt{2}$

8) $30\sqrt{3}$

9) $2\sqrt{6}$

10) $6\sqrt{2}$

11) $11\sqrt{3}$

12) $4 - \sqrt{5}$

13) $4\sqrt{3}$

14) $3 + \sqrt{5}$

15) 6

16) 10

17) 1

18) $4y^3\sqrt{5}$

19) $54\sqrt{a}$

20) 10

21) $6\sqrt{2}$

22) $12\sqrt{3}$

23) $4\sqrt{5}$

24) $8\sqrt{3}$

25) $16\sqrt{5}$

26) 14

Chapter 6:
Ratio, Proportion and Percent

Proportions

Find a missing number in a proportion.

1) $\frac{3}{5} = \frac{18}{a}$

2) $\frac{a}{8} = \frac{25}{40}$

3) $\frac{24}{120} = \frac{a}{10}$

4) $\frac{16}{a} = \frac{96}{36}$

5) $\frac{4}{a} = \frac{16}{75}$

6) $\frac{\sqrt{9}}{4} = \frac{a}{32}$

7) $\frac{2}{4} = \frac{18}{a}$

8) $\frac{7}{14} = \frac{a}{35}$

9) $\frac{7}{a} = \frac{4.2}{6}$

10) $\frac{2}{12} = \frac{8}{a}$

11) $\frac{10}{8} = \frac{5}{a}$

12) $\frac{15}{a} = \frac{3}{13}$

13) $\frac{4}{11} = \frac{a}{12}$

14) $\frac{\sqrt{36}}{3} = \frac{48}{a}$

15) $\frac{6}{a} = \frac{6.6}{39.6}$

16) $\frac{60}{140} = \frac{a}{280}$

17) $\frac{42}{200} = \frac{a}{68}$

18) $\frac{23}{161} = \frac{a}{7}$

19) $\frac{10}{32} = \frac{4}{a}$

20) $\frac{18}{14} = \frac{27}{a}$

Reduce Ratio

Reduce each ratio to the simplest form.

1) 6: 24 =

2) 7: 42 =

3) 72: 40 =

4) 30: 25 =

5) 12: 120 =

6) 16: 2 =

7) 70: 350 =

8) 4: 144 =

9) 25: 75 =

10) 4.8: 5.6 =

11) 110: 330 =

12) 3: 5 =

13) 60: 100 =

14) 24: 36 =

15) 34: 68 =

16) 32: 8 =

17) 140: 35 =

18) 20: 200 =

19) 126: 84 =

20) 156: 198 =

21) 30: 60 =

22) 12: 14 =

23) 10: 150 =

24) 15: 90 =

Percent

Find the Percent of Numbers.

1) 30% of 45 =

2) 23% of 16 =

3) 15% of 17 =

4) 12% of 140 =

5) 7% of 70 =

6) 35% of 12 =

7) 18% of 5 =

8) 12% of 46 =

9) 40% of 62 =

10) 4.5% of 50 =

11) 85% of 18 =

12) 60% of 50 =

13) 18% of 180 =

14) 2% of 240 =

15) 75% of 0 =

16) 80% of 120 =

17) 36% of 45 =

18) 10% of 70 =

19) 8% of 13 =

20) 4% of 8 =

21) 30% of 44 =

22) 80% of 17 =

23) 22% of 35 =

24) 8% of 150 =

25) 40% of 270 =

26) 6% of 15 =

27) 9% of 360 =

28) 10% of 56 =

Discount, Tax and Tip

Find the selling price of each item.

1) Original price of a computer: $300

Tax: 7%, Selling price: $_____

2) Original price of a laptop: $210

Tax: 16%, Selling price: $_____

3) Original price of a sofa: $400

Tax: 8%, Selling price: $_____

4) Original price of a car: $12,600

Tax: 3.5%, Selling price: $_____

5) Original price of a Table: $500

Tax: 12%, Selling price: $_____

6) Original price of a house: $280,000

Tax: 1.5%, Selling price: $_____

7) Original price of a tablet: $460

Discount: 30%, Selling price: $_____

8) Original price of a chair: $110

Discount: 10%, Selling price: $_____

9) Original price of a book: $30

Discount: 10% Selling price: $_____

10) Original price of a cellphone: 720

Discount: 12% Selling price: $_____

11) Food bill: $42

Tip: 10%   Price: $_____

12) Food bill: $38

Tipp: 15%   Price: $_____

13) Food bill: $80

Tip: 27%   Price: $_____

14) Food bill: $62

Tipp: 15%   Price: $_____

Find the answer for each word problem.

15) Nicolas hired a moving company. The company charged $600 for its services, and Nicolas gives the movers a 15% tip. How much does Nicolas tip the movers? $_____

16) Mason has lunch at a restaurant and the cost of his meal is $70. Mason wants to leave a 7% tip. What is Mason's total bill including tip? $_____

Percent of Change

Find each percent of change.

1) From 400 to 800. ___ %

2) From 90 ft to 360 ft. ___ %

3) From $120 to $840. ___ %

4) From 50 cm to 150 cm. ___ %

5) From 10 to 30. ___ %

6) From 60 to 108. ___ %

7) From 120 to 180. ___ %

8) From 400 to 600. ___ %

9) From 170 to 238. ___ %

10) From 200 to 350. ___ %

Calculate each percent of change word problem.

11) Bob got a raise, and his hourly wage increased from $32 to $40. What is the percent increase? ___ %

12) The price of a pair of shoes increases from $50 to $80. What is the percent increase? ___ %

13) At a coffee shop, the price of a cup of coffee increased from $3.50 to $4.2. What is the percent increase in the cost of the coffee? ___ %

14) 22cm are cut from a 88 cm board. What is the percent decrease in length? _ %

15) In a class, the number of students has been increased from 112 to 168. What is the percent increase? ___ %

16) The price of gasoline rose from $36.8 to $42.32 in one month. By what percent did the gas price rise? ___ %

17) A shirt was originally priced at $42. It went on sale for $33.6. What was the percent that the shirt was discounted? ___ %

Simple Interest

Determine the simple interest for these loans.

1) $180 at 15% for 5 years. $ _____
2) $1,600 at 4% for 2 years. $ _____
3) $900 at 25% for 4 years. $ _____
4) $9,200 at 1.5% for 8 months. $ ___
5) $600 at 5% for 7 months. $ _____

6) $40,000 at 8.5% for 3 years. $ ____
7) $7,400 at 8% for 5 years. $ _____
8) $500 at 6.5% for 2 years. $ _____
9) $800 at 4.5 % for 4 months. $ ____
10) $6,000 at 3.5% for 5 years. $ ___

Calculate each simple interest word problem.

11) A new car, valued at $18,000, depreciates at 5.5% per year. What is the value of the car one year after purchase? $_____

12) Sara puts $9,000 into an investment yielding 8% annual simple interest; she left the money in for two years. How much interest does Sara get at the end of those three years? $_____

13) A bank is offering 12.5% simple interest on a savings account. If you deposit $30,400, how much interest will you earn in four years? $_____

14) $1,200 interest is earned on a principal of $5,000 at a simple interest rate of 12% interest per year. For how many years was the principal invested? _____

15) In how many years will $1,800 yield an interest of $432 at 6% simple interest? _____

16) Jim invested $8,000 in a bond at a yearly rate of 2.5%. He earned $600 in interest. How long was the money invested? _____

Answers of Worksheets

Proportions

1) $a = 30$

2) $a = 5$

3) $a = 2$

4) $a = 6$

5) $a = 18.75$

6) $a = 24$

7) $a = 36$

8) $a = 17.5$

9) $a = 10$

10) $a = 48$

11) $a = 4$

12) $a = 65$

13) $a = \frac{48}{11}$

14) $a = 24$

15) $a = 36$

16) $a = 120$

17) $a = 14.28$

18) $a = 1$

19) $a = 12.8$

20) $a = 21$

Reduce Ratio

1) $1: 4$

2) $1: 6$

3) $9: 5$

4) $6: 5$

5) $1: 10$

6) $8: 1$

7) $1: 5$

8) $1: 36$

9) $1: 3$

10) $0.6: 0.7$

11) $11: 33$

12) $0.6: 1$

13) $3: 5$

14) $2: 3$

15) $1: 2$

16) $4: 1$

17) $4: 1$

18) $1: 10$

19) $3: 2$

20) $26: 33$

21) $1: 2$

22) $6: 7$

23) $1: 15$

24) $1: 6$

Percent

1) 13.5

2) 3.68

3) 2.55

4) 16.8

5) 4.9

6) 4.2

7) 0.9

8) 5.52

9) 24.8

10) 2.25

11) 15.3

12) 30

13) 13.4

14) 4.8

15) 0

16) 96

17) 16.2

18) 7

19) 1.04

20) 0.32

21) 13.2

22) 13.6

23) 7.7

24) 12

25) 108

26) 0.9

27) 32.4

28) 5.6

Discount, Tax and Tip

1) $321.00	7) $322.00	13) $101.60
2) $243.6	8) $99.00	14) $71.3
3) $432.00	9) $27.00	15) $90.00
4) $13,041.00	10) $633.60	16) $74.90
5) $560.00	11) $46.2	
6) $284,200	12) $43.7	

Percent of Change

1) 100%	7) 50%	13) 20%
2) 400%	8) 50%	14) 25%
3) 600%	9) 40%	15) 50%
4) 300%	10) 75%	16) 15%
5) 200%	11) 25%	17) 20%
6) 80%	12) 60%	

Simple Interest

1) $135.00	7) $2,960.00	13) $15,200.00
2) $128.00	8) $65.00	14) 2 years
3) $900.00	9) $144.00	15) 4 years
4) $92.00	10) $1,050.00	16) 3 years
5) $17.50	11) $17,010.00	
6) $10,200.00	12) $2,160.00	

Chapter 7:
Monomials and
Polynomials

Adding and Subtracting Monomial

Simplify each expression.

1) $4x^3 + 16x^3 =$

2) $12x^2 + 3x^2 =$

3) $\frac{1}{7}x^3 + \frac{5}{7}x^3 =$

4) $3\frac{2}{8}x^4 + 5\frac{6}{8}x^4 =$

5) $19x^7 - 7x^7 =$

6) $6.9x^3 - 2.9x^3 =$

7) $-x^{10} + x^{10} =$

8) $(x^4)^5 + (x^5)^4 =$

9) $3x^{-5} + 2x^{-5} =$

10) $15p^8 - (-5p^8) =$

11) $2x^2 - 3.8x^2 =$

12) $4\frac{1}{5}x^4 + 3\frac{2}{5}x^4 =$

13) $-4\frac{1}{8}x^{13} + 6\frac{1}{8}x^{13} =$

14) $\sqrt{81}p^6 + (-5p^6) =$

15) $(-1.8p^4) + (-3.2p^4) =$

16) $-1.2x^6 + 7.4x^6 =$

17) $x^8 + \frac{2}{5}x^8 =$

18) $12x^4 - x^4 =$

19) $-3.2x^2 - 6.9x^2 =$

20) $-6x^7 - 3x^7 =$

21) $32x^5 - 20x^5 =$

22) $-12x^{13} + 14x^{13} =$

23) $10x^{-10} - 32x^{-10} =$

24) $27x^{-7} - 17x^{-7} =$

Multiplying and Dividing Monomial

Simplify.

1) $6xy^3 \times 2x^4 =$

2) $9xy \times 4x^3 =$

3) $5xy^3 \times (-4x^3y^4) =$

4) $8x^6y^{10} \times x^3y^2 =$

5) $12x^2 \times (-5x^4) =$

6) $-5x^2y^4z \times 3x^3y^3z^4 =$

7) $-7 \times (-11x^{14}y^{16}) =$

8) $3x^2y^4 \times (-12x^2y^5) =$

9) $6x^2 \times (-8x) =$

10) $-8x^3y^8 \times 3x^2y =$

11) $22x^{-4}y^6 \times (-x^{-7}y^{-3}) =$

12) $6x^{10}y^3z \times 5xy^{-3}z =$

13) $\dfrac{20x^{11}y^4}{10x^5y^2} =$

14) $(6y^5)^{-2} =$

15) $\dfrac{120x^{18}y^6}{12x^8y^3} =$

16) $\dfrac{28x^{14}}{7x^8} =$

17) $\dfrac{33x^7y^4z^3}{11x^2y^4z} =$

18) $\dfrac{15x^2+10x}{5x} =$

19) $\dfrac{200x^5y^9}{100x^5y^8} =$

20) $(12x^2)(6x^2) =$

21) $\dfrac{36x^2y^4+18xy^6}{9xy} =$

22) $\dfrac{-54x^7y^{14}}{6x^5y^{11}} =$

23) $\dfrac{25x^6y^{15}}{10x^4y^4} =$

24) $\dfrac{32x^{16}y^{12}z}{8x^4y^5} =$

Binomial Operations

Solve each operation below.

1) $6x + 16 - (12x - 6) =$

2) $(8x - 10) + (12x - 14) =$

3) $(-5x - 10) + (8x + 4) =$

4) $(4x - 1.4) + (5x - 3.6) =$

5) $\frac{1}{8}x + 4 - \left(\frac{1}{3}x - 5\right) =$

6) $5x + 3 - (7x - 2) =$

7) $14x + 5 - (26x - 1) =$

8) $(x + 8)(x + 5) =$

9) $(x - 7)(x - 3) =$

10) $(x - 4)(3x + 4) =$

11) $(x - 8)(x + 8) =$

12) $(x - 6)(9x + 5) =$

13) $(4x - 5)(4x + 5) =$

14) $(x + 9)(x - 6) =$

15) $(x - 7)(5x + 7) =$

16) $(x^2 + 6)(x^2 - 6) =$

17) $(x - 3)(x + 3) =$

18) $7x(6x - 4) =$

19) $13x(3x + 5) =$

20) $(3x + 4) + (5x - 7) + (x - 9) =$

21) $(x^2 + 2)(x^2 - 2)$

22) $(3x - 3)(5x + 4)$

23) $(x - 5)(7x + 2)$

24) $(x - 3.4)(8.2x + 3.4)$

Polynomial Operations

Simplify each expression.

1) $(8x^2 + 2x - 12) + (4x - 8x^2 - 8) =$

2) $(6x^2 + 4x - 6) - (8x - 6x^2 - 4) =$

3) $(24x^2 - 16x + 6) - (-8x + 12x^2 - 6) =$

4) $(6x^5 - 2x^3 - 7x) + (5x + 14x^4 - 15) + (3x^2 + x^3 + 12) =$

5) $(15x^2 - 8x + 4) + (6x^2 - 2x + 3) =$

6) $12(3x^2 - 7x - 2) =$

7) $3x^3(3x^2 - 3x + 5) =$

8) $3x^2y^2(4x^2 - 6x + 3) =$

9) $(x + 5)(x^2 - 4x + 7) =$

10) $x(3x^2 - 3x + 7) =$

11) $10(x^2 - 3x + 6) =$

12) $(2x - 6)(x^2 + 7x - 2) =$

13) $(12x^3 + 5x^2 - 13) + (-10x^3 + 15x^2 + 22) =$

14) $(6x - 8)(6x^2 + 15x + 20) =$

Squaring a Binomial

Write each square as a trinomial.

1) $(a + 4b)^2 =$

2) $(x + 6)^2 =$

3) $(3a - b)^2 =$

4) $(3x + 2)^2 =$

5) $(x - 8)^2 =$

6) $(2x + \frac{1}{4})^2 =$

7) $(4x - 5y)^2 =$

8) $(x - 7)^2 =$

9) $(x + 11)^2 =$

10) $(4x - 5)^2 =$

11) $(3x + 3y)^2 =$

12) $(3x + 8)^2 =$

13) $(3x + \frac{1}{3})^2 =$

14) $(2x^2 + 2y^2)^2 =$

15) $(x - 12)^2 =$

16) $(x + \sqrt{3})^2 =$

17) $(6x - 7)^2 =$

18) $2(x + 3)^2 =$

19) $(8x - 3y)^2 =$

20) $(x + 13)^2 =$

21) $5(x + 2)^2 =$

22) $(x^2 - 7)^2 =$

23) $(2x + 5)^2 =$

24) $(4x + 6)^2 =$

Factor polynomial

Factor each completely.

1) $x^2 + 17x + 60 =$

2) $11x^2 - 33x =$

3) $x^3 - 6x^2 - 6x + 36 =$

4) $x^2 + 10x + 24 =$

5) $x^4 - 4x^2 - 12 =$

6) $x^2 - 14x + 45 =$

7) $10 + 6x + 32 + x =$

8) $3x^2 - 18x + 16x - 4 =$

9) $24x^3y + 8x^2y - 32xy =$

10) $20x^2 - 7x - 3 =$

11) $3x^2 - 26x + 35 =$

12) $\dfrac{3x^2 - 15x + 18}{x^2 - 9x + 14} =$

13) $\dfrac{(x-3)(x-5)}{(x-3)(x-9)} =$

14) $(x - 4)4x + (x - 4)4 =$

15) $8x^2 - 20x^4 =$

16) $\dfrac{x^2 + 10x + 24}{(x+4)} =$

17) $x^2 + 2x - 63 =$

18) $3x^4 + 21x^2 - 15x^3 - 105x =$

19) $17(a - b) - 5a(a - b) =$

20) $19x^2 - 19x =$

Answers of Worksheets

Adding and Subtracting Monomial

1) $20x^3$
2) $15x^2$
3) $\frac{6}{7}x^3$
4) $9x^4$
5) $12x^7$
6) $4x^3$
7) 0
8) $2x^{20}$

9) $5x^{-5}$
10) $20p^8$
11) $-1.8x^2$
12) $7\frac{3}{5}x^4$
13) $2x^{13}$
14) $4p^6$
15) $-5p^4$
16) $6.2x^6$

17) $\frac{7}{5}x^8$
18) $11x^4$
19) $-10.1x^2$
20) $-9x^7$
21) $12x^5$
22) $2x^{13}$
23) $-22x^{-10}$
24) $10x^{-7}$

Multiplying and Dividing Monomial

1) $12x^5y^3$
2) $36x^4y$
3) $-20x^4y^7$
4) $8x^9y^{12}$
5) $-60x^6$
6) $-15x^5y^7z^5$
7) $77x^{14}y^{16}$
8) $-36x^4y^9$
9) $-48x^3$

10) $-24x^5y^9$
11) $-22x^{-11}y^3$
12) $30x^{11}z^2$
13) $2x^6y^2$
14) $\frac{1}{36}y^{-10}$
15) $10x^{10}y^3$
16) $4x^6$
17) $3x^5z^4$

18) $3x + 2$
19) $2y$
20) $72x^4$
21) $4xy^3 + 2y^5$
22) $-9x^2y^3$
23) $\frac{5}{2}x^2y^{11}$
24) $4x^{12}y^7z$

Binomial Operations

1) $-6x + 22$
2) $20x - 24$
3) $3x - 6$
4) $9x - 5$
5) $-\frac{5}{24}x + 9$
6) $-2x + 5$
7) $-12x + 6$
8) $x^2 + 13x + 40$
9) $x^2 - 10x + 21$

10) $3x^2 - 8x - 16$
11) $x^2 - 64$
12) $9x^2 - 49x - 30$
13) $16x^2 - 25$
14) $x^2 + 3x - 54$
15) $5x^2 - 28x - 49$
16) $x^4 - 36$
17) $x^2 - 9$
18) $42x^2 - 28x$

19) $39x^2 + 65x$
20) $9x - 12$
21) $x^4 - 4$
22) $15x^2 - 3x - 12$
23) $7x^2 - 33x - 10$
24) $8.2x^2 - 24.48x - 11.56$

Polynomial Operations

1) $6x - 20$

2) $12x^2 - 4x - 2$

3) $12x^2 - 8x + 12$

4) $6x^5 + 14x^4 - x^3 + 3x^2 - 3$

5) $21x^2 - 10x + 7$

6) $36x^2 - 84x - 24$

7) $9x^5 - 9x^4 + 15x^3$

8) $12x^4y^2 - 18x^3y^2 + 9x^2y^2$

9) $x^3 + x^2 - 13x + 35$

10) $3x^3 - 3x^2 + 7x$

11) $10x^2 - 30x + 60$

12) $2x^3 + 8x^2 - 46x + 12$

13) $2x^3 + 20x^2 + 9$

14) $36x^3 + 42x^2 - 160$

Squaring a Binomial

1) $a^2 + 16b^2 + 8ab$

2) $x^2 + 12x + 36$

3) $9a^2 + b^2 - 6ab$

4) $9x^2 + 12x + 4$

5) $x^2 - 16x + 64$

6) $x^2 + x + \frac{1}{16}$

7) $16x^2 - 40xy + 25y^2$

8) $x^2 - 14x + 49$

9) $x^2 + 22x + 121$

10) $16x^2 - 40x + 25$

11) $9x^2 + 18xy + 9y^2$

12) $9x^2 + 48x + 64$

13) $9x^2 + 2x + \frac{1}{9}$

14) $4x^4 + 4y^4 + 8x^2y^2$

15) $x^2 - 24x + 144$

16) $x^2 + 2\sqrt{3}x + 3$

17) $36x^2 - 84x + 49$

18) $2x^2 + 12x + 18$

19) $64x^2 - 48xy + 9y^2$

20) $x^2 + 36x + 169$

21) $5x^2 + 20x + 20$

22) $x^4 - 14x^2 + 49$

23) $4x^2 + 20x + 25$

24) $16x^2 + 48x + 36$

Factor polynomial

1) $(x + 5)(x + 12)$

2) $11x(x - 3)$

3) $(x^2 - 6)(x - 6)$

4) $(x + 6)(x + 4)$

5) $(x^2 - 6)(x^2 + 2)$

6) $(x - 9)(x - 5)$

7) $7(x + 6)$

8) $3x(x - 2) + 4(x - 1)$

9) $8xy(3x^2 + x - 4)$

10) $(4x + 1)(5x - 3)$

11) $(3x - 5)(x - 7)$

12) $\frac{3(x-3)}{x-7}$

13) $\frac{x-5}{x-9}$

14) $(x - 4)(4x + 4)$

15) $-2x^2(-4 + 10x^2)$

16) $x + 6$

17) $(x + 9)(x - 7)$

18) $3x(x^2 + 7)(x - 5)$

19) $(a - b)(17 - 5a)$

20) $19x(x - 1)$

Chapter 8:

Functions

Relation and Functions

Determine whether each relation is a function. Then state the domain and range of each relation.

1)

Function:

....

Domain:

....

Range:

....

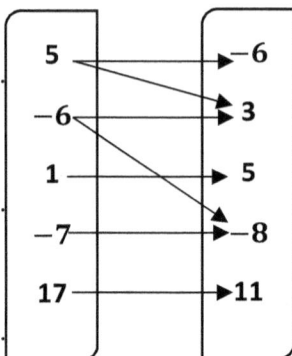

2)

x	y
7	6
3	4
−8	−9
8	−9
−11	2

Function:

....

Domain:

....

Range:

....

3)

Function:

....

Domain:

....

Range:

....

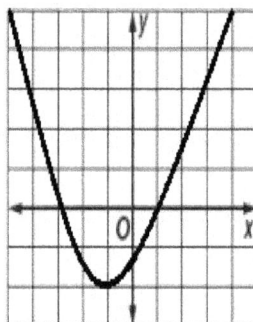

4) $\{(2, -2), (7, -6), (9, 9), (8,1), (7,4)\}$

Function:

....

Domain:

....

Range:

....

5)

Function:

....

Domain:

....

Range:

....

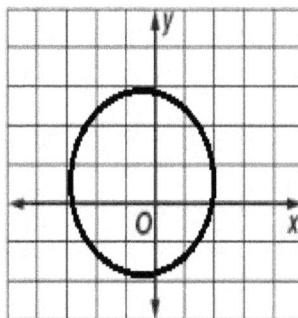

6)

Function:

....

Domain:

....

Range:

....

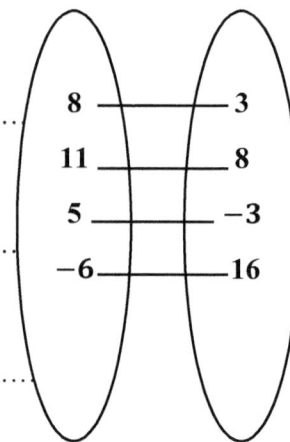

Slope form

Write the slope-intercept form of the equation of each line.

1) $6x + 7y = 14$

2) $8x + 24y = 6$

3) $14x + 2y = -18$

4) $-7x + 11y = 5$

5) $5x - 4y = 7$

6) $-21x + 3y = 6$

7) $2x + y = 0$

8) $5x - 7y = -9$

9) $-13.5x + 27y = 54$

10) $-3x + \frac{2}{3}y = 18$

11) $10x + y = -120$

12) $6x = -72y - 54$

13) $4.5x = 9y + 9$

14) $10x = -\frac{10}{8}y + 50$

Slope and Y-Intercept

Find the slope and y-intercept of each equation.

1) $y = \frac{1}{8}x + 5$

2) $y = 10x + 7$

3) $x - 6y = 18$

4) $y = 10x + 22$

5) $y = 9$

6) $y = -2x + 3$

7) $x = -15$

8) $y = 7x$

9) $y - 5 = 8(x + 1)$

10) $x = -\frac{11}{8}y - \frac{1}{6}$

Slope and One Point

Find a Point-Slope equation for a line containing the given point and having the given slope.

1) $m = -2, (1, -1)$

2) $m = 3, (1, 2)$

3) $m = -2, (-1, -5)$

4) $m = 2, (6, 4)$

5) $m = 5, (2, 4)$

6) $m = \frac{3}{2}, (4, 5)$

7) $m = 0, (-4, -5)$

8) $m = 2, (1, -3)$

9) $m = 1, (0, 3)$

10) $m = \frac{3}{4}, (-2, -5)$

11) $m = -3, (1, -1)$

12) $m = -2, (2, -1)$

13) $m = 5, (1, 0)$

14) $m =$ undefined, $(8, -8)$

15) $m = -\frac{1}{8}, (8, 4)$

16) $m = \frac{1}{4}, (3, 2)$

17) $m = -8, (2, 4)$

18) $m = 6, (-2, -4)$

19) $m = \frac{1}{3}, (3, 1)$

20) $m = \frac{-4}{9}, (0, -3)$

21) $m = \frac{1}{4}, (4, 4)$

22) $m = -5, (0, -1)$

23) $m = 0, (0.9, -3)$

24) $m = -\frac{2}{5}, (5, -2)$

25) $m = 0, (-2, 17)$

26) $m =$ Undefined, $(-9, -3)$

Slope of Two Points

Write the slope-intercept form of the equation of the line through the given points.

1) $(2, 0), (-2, 10)$

2) $(-2, 6), (10, 12)$

3) $(-10, 2), (-2, 10)$

4) $(2, -3), (-9, 8)$

5) $(5, 0), (3, 1)$

6) $(9, -1), (-1, 9)$

7) $(-5, 3), (-6, 1)$

8) $(-7, -2), (1, 0)$

9) $(-5, -5), (3, 3)$

10) $(-1, 9), (-1, -5)$

11) $(-2, 7), (1, 7)$

12) $(1, -5), (4, -4)$

13) $(6, -9), (-3, 0)$

14) $(1, -4), (7, 4)$

15) $(-9, 5), (-3, -1)$

16) $(9, 5), (5, 1)$

17) $(10, -7), (2, -6)$

18) $(-5, -9), (-7, 2)$

19) $(7, 4), (3, 1)$

20) $(-1, -1), (9, 2)$

21) $(-8, 8), (8, 2)$

22) $(9, 2), (5, 11)$

23) $(16, 4), (18, 6)$

24) $(-4, -10), (-10, -16)$

Equation of Parallel and Perpendicular lines

Write the slope-intercept form of the equation of the line described.

1) Through: $(-5, 4)$, parallel to $y = 4x + 10$

2) Through: $(-2, 1)$, parallel to $y = -7x$

3) Through: $(-10, -2)$, perpendecular to $y = \frac{1}{2}x + 8$

4) Through: $(6, -2)$, parallel to $y = -5x + 13$

5) Through: $(-7, 4)$, parallel to $y = \frac{3}{7}x - 6$

6) Through: $(2, 0)$, perpendecular to $y = -\frac{1}{5}x + 8$

7) Through: $(4, -7)$, perpendecular to $y = -6x - 10$

8) Through: $(-5, 1)$, perpendecular to $y = -\frac{1}{8}x + 3$

9) Through: $(-1, -2)$, parallel to $2y + 4x = 9$

10) Through: $(1, 10)$, parallel to $y = \frac{1}{10}x - 5$

11) Through: $(5, -5)$, parallel to $y = 9$

12) Through: $(7, 2)$, perpendecular to $y = \frac{5}{2}x + 3$

13) Through: $(0, -4)$, perpendecular to $3y - x = 11$

14) Through: $(3, 5)$, parallel to $3y + x = 5\frac{3}{4}$

15) Through: $(1, 1)$, perpendecular to $y = 5x + 12$

16) Through: $(-2, -4)$, parallel to $6y - x = 11$

17) Through: $(-1, -1)$, perpendecular to $y = x + \frac{1}{2}$

18) Through: $(-6, 0)$, perpendecular to $5y - 2x - 9 = 0$

Quadratic Equations - Square Roots Law

Solve each equation by taking square roots.

1) $x^2 - 3 = 6$

2) $2x^2 - 12 = 38$

3) $24x^2 - 4 = 200$

4) $-x^2 - 2 = -66$

5) $6x^2 + 3 = 489$

6) $5x^2 + 9 = 14$

7) $8x^2 - 17 = 2,791$

8) $7x^2 + 16 = 2,151$

9) $100x^2 = 4$

10) $4x^2 - 8 = 68$

11) $9x^2 - 5 = 607$

12) $20x^2 - 20 = 60$

13) $13x^2 - 3 = 4,209$

14) $13x^2 - 8 = -1,139$

15) $16x^2 - 16 = 48$

16) $54x^2 = 6$

17) $-8x^2 - 8 = -31$

18) $15x^2 - 30 = 255$

19) $42x^2 = -126$

20) $16x^2 - 10 = 39$

21) $2x^2 + 16 = 160$

22) $13x^2 = 117$

23) $-16x^2 + 12 = 340$

24) $6x^2 - 30 = -24$

25) $12x^2 - 60 = -48$

26) $6x^2 + 30 = 630$

27) $54x^2 + 6 = 1,950$

28) $42x^2 + 6 = 174$

Quadratic Equations - Factoring

Solve each equation by factoring.

1) $(20n - 10)(18n + 6) = 0$

2) $(40n - 8)(8n + 8) = 0$

3) $2x^2 + 32 = 20x$

4) $7x^2 - 42 = -35x$

5) $2x^2 - 7 = 13x$

6) $7x^2 + 32 = 7 - 40x$

7) $-6x^2 + 2x + 144 = 6x^2 + 14x$

8) $8x^2 + 3x + 2 = 7x^2$

9) $8x^2 - 57x = -7$

10) $21x^2 + 75 = -120x$

11) $3n(2n - 10) = 0$

12) $(n + 4)(5n - 6) = 0$

13) $x^2 + 9 = -9 - 9x$

14) $5x^2 + 12 = -20x - 8$

15) $3x^2 + 20 = -18x - 7$

16) $5x^2 + 2x = 10 - 3x$

17) $3x^2 - 8x + 4 = 20$

18) $2x^2 + 40 = -24x - 32$

19) $3x^2 - 8x = 16$

20) $3x^2 + 50 = -30x - 25$

21) $6x^2 - 16x - 32 = 0$

22) $10x^2 + 10x - 100 = 20$

23) $4x^2 + 12x = -20x - 64$

24) $4n(2n + 4) = 0$

Quadratic Equations - Completing the Square

Solve each equation by completing the square.

1) $12x^2 - 24x - 36 = 0$

2) $20x^2 - 40x - 60 = 0$

3) $4x^2 + \frac{10x}{2} - 6 = 0$

4) $-3x^2 + 2x + 8 = 0$

5) $3x^2 + 42x - 153 = 0$

6) $3x^2 + 18x + 4 = -20$

7) $x^2 + 14x - 2 = 13$

8) $x^2 - 4x - 91 = 7$

9) $2x^2 - 18x = -36$

10) $5x^2 = -20x + 60$

11) $12x^2 = 48x - 36$

12) $\frac{1}{4}x^2 - \frac{2}{4}x - \frac{3}{4} = 0$

13) $3x^2 - 36x + 25 = -8$

14) $3x^2 - 30x + 54 = 0$

15) $x^2 + 6x = 59$

16) $3x^2 = 54x + 120$

17) $\frac{1}{5}x^2 + \frac{2}{5}x = -4$

18) $2x^2 - 24x = 22$

19) $6x^2 - 12x - 18 = 0$

20) $3x^2 + 42x - 45 = 0$

21) $\frac{1}{2}x^2 - x - \frac{3}{2} = 0$

22) $14x^2 - 56x = -42$

23) $18x^2 + 108x + 144 = 0$

24) $-14x^2 - 84x = 112$

Quadratic Equations - Quadratic Formula

Solve each equation with the quadratic formula.

1) $6x^2 + 12x - 48 = 0$

2) $8x^2 - 16x = -8$

3) $\frac{1}{8}x^2 = \frac{9}{8}x - \frac{5}{2}$

4) $\frac{2}{3}x^2 + \frac{10}{3}x - 4 = 0$

5) $3x^2 = 27x - 60$

6) $3x^2 - 12x - 26 = 10$

7) $7x^2 = -21x + 280$

8) $3x^2 + 15x - 18 = 0$

9) $x^2 + x - 2 = \frac{1}{4}$

10) $\frac{4}{3}x^2 - \frac{2}{3}x = 3$

11) $x^2 = -3x + 40$

12) $9x^2 - 25 = 8x$

13) $\frac{8}{5}x^2 - \frac{8}{5}x = 3$

14) $6x^2 - 3x - 10 = 8$

15) $\frac{1}{3}x^2 = 3x - \frac{100}{15}$

16) $10x^2 + 30x = 400$

17) $x^2 - 2 = \frac{1}{8}x$

18) $24x^2 + 18x + 15 = 0$

19) $x^2 - \frac{1}{2}x - \frac{13}{2} = 1$

20) $22x^2 + 2 = 10x^2 + 14x$

21) $34x^2 + 28 = 2x$

22) $20x^2 - 10x - 40 = 20$

Arithmetic Sequences

Find the three terms in the sequence after the last one given.

1) $3, 9, 15, 21, ..., ..., ...,$

2) $-12, -4, 4, 12, ..., ..., ...,$

3) $a_1 = -10, d = 5$

4) $a_1 = -11.2, d = 1.5$

5) $a_{10} = 36, d = 6$

6) $a_n = (3n)^2$

7) $24, 16, 8, 0, ..., ..., ...,$

8) $-15, -30, -45, -60, ..., ..., ...,$

9) $a_n = -18 + 3n$, find a_{26}

10) $a_n = -8.4 - 5.8n$, find a_{32}

11) $a_5 = \frac{3}{8}, d = -\frac{1}{4}$

12) $a_n = \frac{2n^2}{3n+2}$

13) $-8, -15, -22, ..., ..., ...,$

14) $a_n = 4n + 3$

15) $-8, -3, 2, 7, ..., ..., ...,$

16) $-12, -7, -2, 3, ..., ..., ...,$

17) $\frac{6}{7}, \frac{5}{14}, -\frac{1}{7}, -\frac{9}{14}, ..., ..., ...,$

18) $5, -5, -15, ..., ..., ...,$

19) $9, 13, 17, ..., ..., ...,$

20) $11, 22, 33, ..., ..., ...,$

21) $-3.5, -4.7, -5.9, ..., ..., ...,$

22) $115, 140, 165, 190, ..., ..., ...,$

23) $-\frac{8}{9}, -\frac{5}{9}, -\frac{2}{9}, ..., ..., ...,$

24) $-3.05, -2.55, -2.05, ..., ..., ...,$

Geometric Sequences

Find the three terms in the sequence after the last one given.

1) $2, 4, 8, 16, \ldots, \ldots, \ldots$

2) $250, 125, 62.5, \ldots, \ldots, \ldots$

3) $a_n = a_{n-1} \times (-2), a_1 = 1$

4) $a_n = a_{n-1} \times 4, a_1 = 3$

5) $0.2, 0.6, 1.8, 5.4 \ldots, \ldots, \ldots$

6) $a_n = 2^{n-1}, a_1 = 1$

7) $a_n = 8(\frac{1}{4})^{n-1}, a_1 = 8$

8) $1, \frac{3}{2}, \frac{9}{4}, \frac{27}{8}, \ldots \ldots, \ldots, \ldots$

9) $a_n = 2a_{n-1}, \ a_1 = 5$

10) $1, 2, 4, \ldots, \ldots, \ldots$

11) $-4, -12, -36, \ldots, \ldots, \ldots$

12) $-0.2, 1, -5, \ldots, \ldots, \ldots$

13) $a_n = (-2)^{2n+1}$

14) $a_n = 0.25 \times (4)^{n-1} \ find \ a_5$

15) $a_n = 3 \times (2)^{n-1} \ ; \ a_6 = ?$

16) $3, 1, \frac{1}{3}, \ldots, \ldots, \ldots, \ldots$

17) $2, 6, 18, \ldots, \ldots, \ldots$

18) $5, -5, 5, \ldots, \ldots, \ldots$

19) $243, 162, 108, \ldots, \ldots, \ldots$

20) $-1.5, 3, -6, \ldots, \ldots, \ldots$

21) $-0.25, -1, -4, \ldots, \ldots, \ldots \ldots$

22) $1, -4, 16, \ldots, \ldots, \ldots$

23) $a_n = -0.1(-2)^{n-1}$

24) $a_n = -0.5 \times 4^{n-1}$

Answers of Worksheets

Relation and Functions

1) No, $D_f = \{5, -6, 1, -7, 17\}$, $R_f = \{-6, 3, 5, -8, 11\}$

2) Yes, $D_f = \{7, 3, -8, 8, -11\}$, $R_f = \{6, 4, -9, 2\}$

3) Yes, $D_f = (-\infty, \infty)$, $R_f = \{-2, \infty)$

4) No, $D_f = \{2, 7, 9, 8, 7\}$, $R_f = \{-2, -6, 9, 1, 4\}$

5) No, $D_f = [-3, 2]$, $R_f = [-2, 3]$

6) Yes, $D_f = \{8, 11, 5, -6\}$, $R_f = \{2, 8, -3, 16\}$

Slope form

1) $y = -\frac{6}{7}x + 2$

2) $y = -\frac{1}{3}x + \frac{1}{4}$

3) $y = -7x - 9$

4) $y = \frac{7}{11}x + \frac{5}{11}$

5) $y = \frac{5}{4}x - \frac{7}{4}$

6) $y = 7x + 2$

7) $y = -2x$

8) $y = \frac{5}{7}x + \frac{9}{7}$

9) $y = 0.5x + 2$

10) $y = 4.5x + 27$

11) $y = -10x - 120$

12) $y = -\frac{1}{12}x - \frac{3}{4}$

13) $y = 0.5x - 1$

14) $y = -8x + 40$

Slope and Y-Intercept

1) $m = \frac{1}{8}, b = 5$

2) $m = 10, b = 7$

3) $m = \frac{1}{6}, b = -3$

4) $m = 10, b = 22$

5) $m = 0, b = 9$

6) $m = -2, b = 3$

7) $m = undefind,$
   $b: no\ intercept$

8) $m = 7, b = 0$

9) $m = 8, b = 13$

10) $m = -\frac{8}{11}, b = -\frac{1}{6}$

Slope and One Point

1) $y = -2x + 1$

2) $y = 3x - 1$

3) $y = -2x - 7$

4) $y = 2x - 8$

5) $y = 5x - 6$

6) $y = \frac{3}{2}x - 1$

7) $y = -5$

8) $y = 2x - 5$

9) $y = x + 3$

10) $y = \frac{3}{4}x - \frac{7}{2}$

11) $y = -3x + 2$

12) $y = -2x + 3$

13) $y = 5x$

14) $x = 8$

15) $y = -\frac{1}{8}x + 5$

16) $y = \frac{1}{4}x + \frac{5}{4}$

17) $y = -8x + 20$

18) $y = 6x + 8$

19) $y = \frac{1}{3}x$

20) $y = -\frac{4}{9}x - 3$

21) $y = \frac{1}{4}x + 3$

22) $y = -5x - 1$

23) $y = -3$

24) $y = -\frac{2}{5}x$

25) $y = 17$

26) $x = -9$

Slope of Two Points

1) $y = -\frac{5}{2}x + 5$

2) $y = \frac{1}{2}x + 7$

3) $y = x + 12$

4) $y = -x - 1$

5) $y = -\frac{1}{2}x + \frac{5}{2}$

6) $y = -x + 8$

7) $y = 2x + 13$

8) $y = \frac{1}{4}x - \frac{1}{4}$

9) $y = x$

10) $x = -1$

11) $y = 7$

12) $y = \frac{1}{3}x - 5\frac{1}{3}$

13) $y = -x - 3$

14) $y = \frac{4}{3}x - 5\frac{1}{3}$

15) $y = -x - 4$

16) $y = x - 4$

17) $y = -\frac{1}{8}x - 5\frac{3}{4}$

18) $y = -5\frac{1}{2}x - 36\frac{1}{2}$

19) $y = \frac{3}{4}x - 1\frac{1}{4}$

20) $y = \frac{3}{10}x - \frac{7}{10}$

21) $y = -\frac{3}{8}x + 5$

22) $y = -\frac{9}{4}x + 22\frac{1}{4}$

23) $y = x - 12$

24) $y = x - 6$

Equation of Parallel and Perpendicular lines

1) $y = 4x + 24$

2) $y = -7x - 13$

3) $y = -2x - 22$

4) $y = -5x + 28$

5) $y = \frac{3}{7}x + 7$

6) $y = 5x - 10$

7) $y = \frac{1}{6}x - 7\frac{2}{3}$

8) $y = 8x + 41$

9) $y = -2x - 4$

10) $y = \frac{1}{10}x + 9\frac{9}{10}$

11) $y = -5$

12) $y = -\frac{2}{5}x + 4\frac{4}{5}$

13) $y = -3x - 4$

14) $y = -\frac{1}{3}x + 6$

15) $y = -\frac{1}{5}x + 1\frac{1}{5}$

16) $y = \frac{1}{6}x - 3\frac{2}{3}$

17) $y = -x - 2$

18) $y = -\frac{5}{2}x - 15$

Quadratic Equations - Square Roots Law

1) $\{3, -3\}$

2) $\{5, -5\}$

3) $\{\frac{\sqrt{17}}{2}, -\frac{\sqrt{17}}{2}\}$

4) $\{8, -8\}$

5) $\{9, -9\}$

6) $\{1, -1\}$

7) $\{3\sqrt{39}, -3\sqrt{39}\}$

8) $\{\sqrt{305}, -\sqrt{305}\}$

9) $\{\frac{1}{5}, -\frac{1}{5}\}$

10) $\{\sqrt{19}, -\sqrt{19}\}$

11) $\{2\sqrt{17}, -2\sqrt{17}\}$

12) $\{2, -2\}$

13) $\{18, -18\}$

14) $\{i\sqrt{87}, -i\sqrt{87}\}$

15) $\{2, -2\}$

16) $\{\frac{1}{3}, -\frac{1}{3}\}$

17) $\{\frac{\sqrt{46}}{4}, -\frac{\sqrt{46}}{4}\}$

18) $\{\sqrt{19}, -\sqrt{19}\}$

19) $\{i\sqrt{3}, -i\sqrt{3}\}$

20) $\{\frac{7}{4}, -\frac{7}{4}\}$

21) $\{6\sqrt{2}, -6\sqrt{2}\}$

22) $\{3, -3\}$

23) $\{i\sqrt{\frac{41}{2}}, -i\sqrt{\frac{41}{2}}\}$

24) $\{1, -1\}$

25) $\{1, -1\}$

26) $\{10, -10\}$

27) $\{6, -6\}$

28) $\{2, -2\}$

Quadratic Equations - Factoring

1) $\{\frac{1}{2}, -\frac{1}{3}\}$

2) $\{\frac{1}{5}, -1\}$

3) $\{2, 8\}$

4) $\{-6, 1\}$

5) $\{-\frac{1}{2}, 7\}$

6) $\{-\frac{5}{7}, -5\}$

7) $\{3, -4\}$

8) $\{-2, -1\}$

9) $\{\frac{1}{8}, 7\}$

10) $\{-\frac{5}{7}, -5\}$

11) $\{5, 0\}$

12) $\{-4, \frac{6}{5}\}$

13) $\{-6, -3\}$

14) $\{-2\}$

15) $\{-3\}$

16) $\{-2, 1\}$

17) $\{-\frac{4}{3}, 4\}$

18) $\{-6\}$

19) $\{-\frac{4}{3}, 4\}$

20) $\{-5\}$

21) $\{-\frac{4}{3}, 4\}$

22) $\{-4, 3\}$

23) $\{-4\}$

24) $\{-2, 0\}$

Quadratic Equations - Completing the Square

1) $\{-1, 3\}$

2) $\{3, -1\}$

3) $\{-2, \frac{3}{4}\}$

4) $\{2, -\frac{4}{3}\}$

5) $\{-17, 3\}$

6) $\{-2, -4\}$

7) $\{-15, 1\}$

8) $\{2+\sqrt{102}, 2-\sqrt{102}\}$

9) $\{3, 6\}$

10) $\{-6, 2\}$

11) $\{3, 1\}$

12) $\{-1, 3\}$

13) $\{11, 1\}$

14) $\{5+\sqrt{7}, 5-\sqrt{7}\}$

15) $\{-3+2\sqrt{17}, -3-2\sqrt{17}\}$

16) $\{-2, 20\}$

17) $\{-1+i\sqrt{19}, -1-i\sqrt{19}\}$

18) $\{6+\sqrt{47}, 6-\sqrt{47}\}$

19) $\{3, -1\}$

20) $\{-15, 1\}$

21) $\{-1, 3\}$

22) $\{3, 1\}$

23) $\{-2, -4\}$

24) $\{-2, -4\}$

Quadratic Equations - Quadratic Formula

1) $\{2, -4\}$

2) $\{1\}$

3) $\{5, 4\}$

4) $\{1, -6\}$

5) $\{5, 4\}$

6) $\{6, -2\}$

7) $\{5, -8\}$

8) $\{1, -6\}$

9) $\{\frac{-1+\sqrt{10}}{2}, \frac{-1-\sqrt{10}}{2}\}$

10) $\{\frac{1+\sqrt{37}}{4}, \frac{1-\sqrt{37}}{4}\}$

11) $\{5, -8\}$

12) $\{\frac{4+\sqrt{241}}{9}, \frac{4-\sqrt{241}}{9}\}$

13) $\{\frac{2+\sqrt{34}}{4}, \frac{2-\sqrt{34}}{4}\}$

14) $\{2, -\frac{3}{2}\}$

15) $\{5, 4\}$

16) $\{5, -8\}$

17) $\left\{ \frac{1+3\sqrt{57}}{16}, \frac{1-3\sqrt{57}}{16} \right\}$

18) $\left\{ \frac{-3+i\sqrt{31}}{8}, \frac{-3-i\sqrt{31}}{8} \right\}$

19) $\left\{ 3, -\frac{5}{2} \right\}$

20) $\left\{ 1, \frac{1}{6} \right\}$

21) $\left\{ \frac{1+i\sqrt{951}}{34}, \frac{1-i\sqrt{951}}{34} \right\}$

22) $\left\{ 2, -\frac{3}{2} \right\}$

Arithmetic sequences

1) $3, 9, 15, 21, 27, 33, 39$

2) $-12, -4, 4, 12, 20, 28, 36$

3) $-10, -5, 0, 5$

4) $-11.2, -9.7, -8.2, -6.7$

5) $-18, -12, -6, 0$

6) $9, 36, 81, 144$

7) $24, 16, 8, 0, -8, -16, -24$

8) $-15, -30, -45, -60, -75, -90, -105$

9) 60

10) -194

11) $\frac{11}{8}, \frac{9}{8}, \frac{7}{8}, \frac{5}{8}$

12) $\frac{2}{5}, 1, \frac{18}{11}, \frac{32}{14}$

13) $-8, -15, -22, -29, -36, -43$

14) $7, 11, 15, 19$

15) $-8, -3, 2, 7, 12, 17, 22$

16) $-12, -7, -2, 3, 8, 13, 18$

17) $\frac{6}{7}, \frac{5}{14}, -\frac{1}{7}, -\frac{9}{14}, -\frac{8}{7}, -\frac{23}{14}, -\frac{15}{7}$

18) $5, -5, -15, -25, -35, -45$

19) $9, 13, 17, 21, 25, 29$

20) $11, 22, 33, 44, 55, 66$

21) $-3.5, -4.7, -5.9 - 7.1, -8.3, -9.5$

22) $115, 140, 165, 190, 215, 240, 265$

23) $-\frac{8}{9}, -\frac{5}{9}, -\frac{2}{9}, \frac{1}{9}, \frac{4}{9}, \frac{7}{9}$

24) $-3.05, -2.55, -2.05, -1.55, -1.05$

Geometric sequences

1) $2, 4, , 8, 16, 32, 64, 128$

2) $250, 125, 62.5, 31.25, 15.625, 7.8125$

3) $1, -2, 4, -8$

4) $3, 12, 48, 192$

5) $0.2, 0.6, 1.8, 5.4, 16.2, 48.6, 145.8$

6) $1, 2, 4, 8$

7) $8, 2, \frac{1}{2}, \frac{1}{8}$

8) $\frac{81}{16}, \frac{243}{32}, \frac{729}{64}$

9) $5, 10, 20, 40$

10) $1, 2, 4, 8, 16, 32$

11) $-4, -12, -36, -108, -324, -972$

12) $-0.2, 1, -5, 25, -125, 625$

13) $-8, -32, -128$

14) 64

15) 96

16) $3, 1, \frac{1}{3}, \frac{1}{9}, \frac{1}{27}, \frac{1}{81}$

17) $2, 6, 18, 54, 162, 486$

18) $5, -5, 5, -5, 5, -5$

19) $243, 162, 108, 72, 48, 32$

20) $-1.5, 3, -6, 12, -24, 48$

21) $-0.25, -1, -4, -16, -64, 256$

22) $1, -4, 16, -64, 256, -1,024$

23) $-0.1, \ 0.2, -0.4, 0.8$

24) $-0.5, -2, -8, -32$

Chapter 9:

Geometry

Area and Perimeter of Square

Find the perimeter and area of each squares.

1)

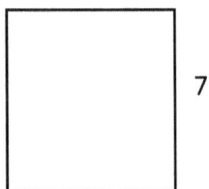

Perimeter:_____:

Area:_____:

2)

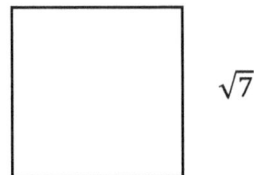

Perimeter:_____:

Area:_____:

3)

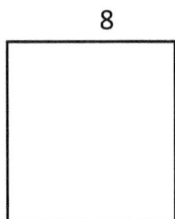

Perimeter:_____:

Area:_____:

4)

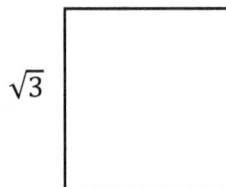

Perimeter:_____:

Area:_____:

5)

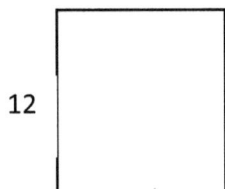

Perimeter:_____:

Area:_____:

6)

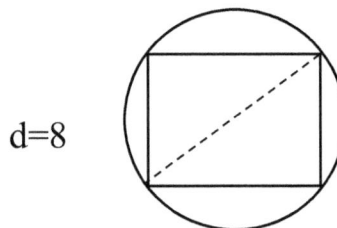

Perimeter of Square:_____:

Area of Square:_____:

Area and Perimeter of Rectangle

Find the perimeter and area of each rectangle.

1)

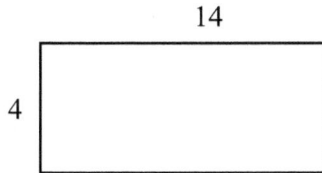

Perimeter:_____:

Area:_____:

2)

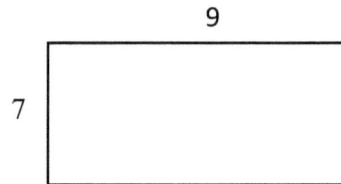

Perimeter:_____:

Area:_____:

3)

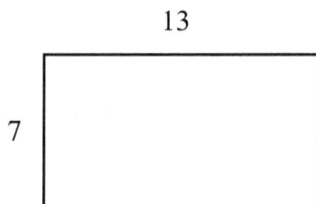

Perimeter:_____:

Area:_____:

4)

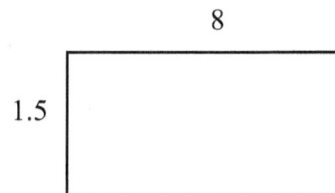

Perimeter:_____:

Area:_____:

5)

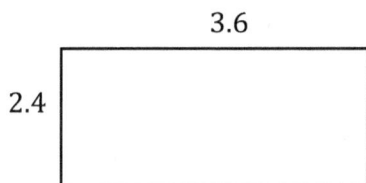

Perimeter:_____:

Area:_____:

6)

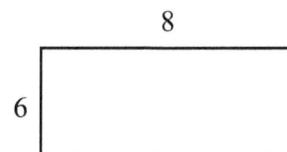

Perimeter:_____:

Area:_____:

Area and Perimeter of Triangle

Find the perimeter and area of each triangle.

1)

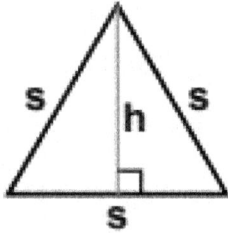

Perimeter:_____.

Area:_____.

2)

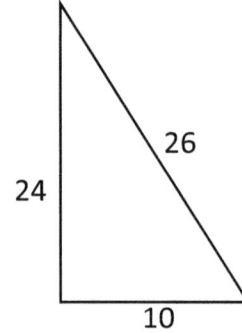

Perimeter:_____.

Area:_____.

3)

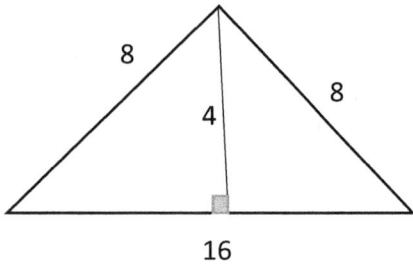

Perimeter:_____.

Area _____:

4)

s=12

h=8

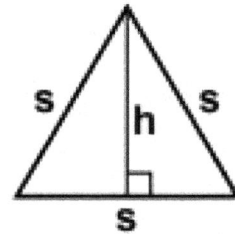

Perimeter:_____.

Area:_____.

5)

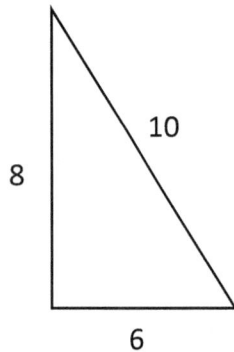

Perimeter:_____.

Area:_____.

6)

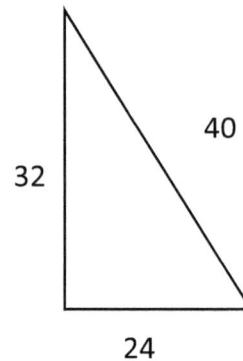

Perimeter:_____.

Area:_____.

Area and Perimeter of Trapezoid

Find the perimeter and area of each trapezoid.

1)

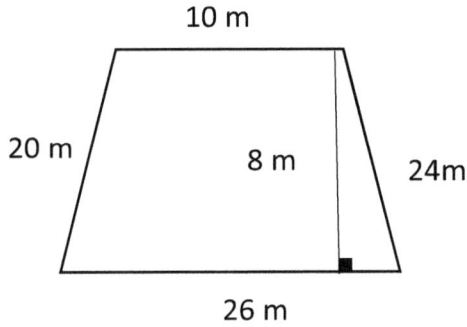

Perimeter:_____:

Area:_____:

2)

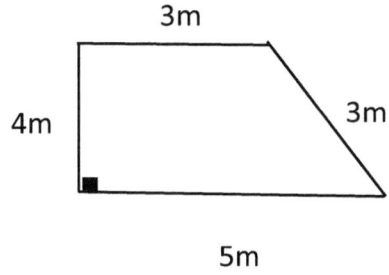

Perimeter:_____:

Area:_____:

3)

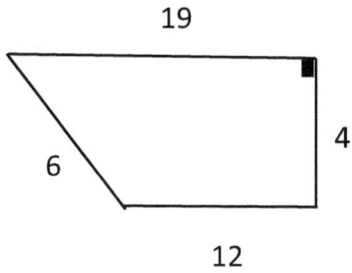

Perimeter:_____:

Area:_____:

4)

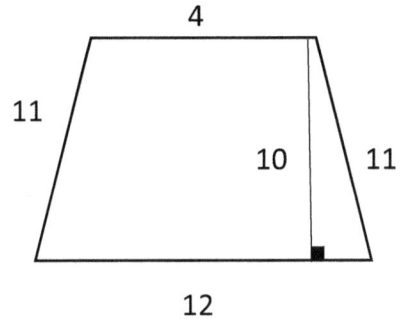

Perimeter:_____:

Area:_____:

5)

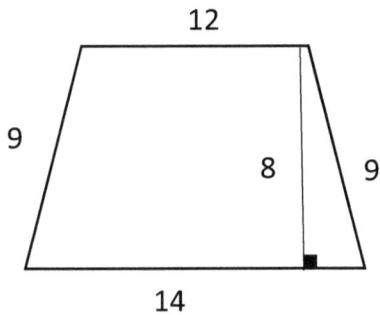

Perimeter:_____:

Area:_____:

6)

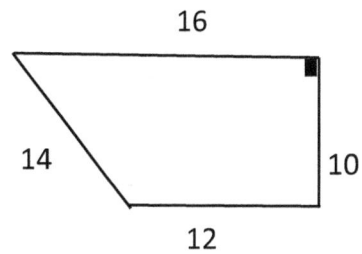

Perimeter:_____:

Area:_____:

Area and Perimeter of Parallelogram

Find the perimeter and area of each parallelogram.

1)

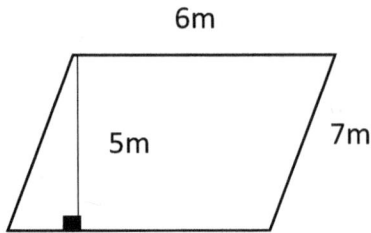

Perimeter:_____.

Area:_____.

2)

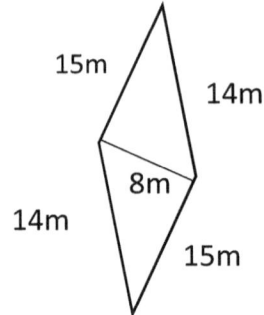

Perimeter:_____.

Area:_____.

3)

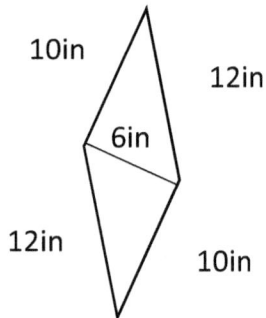

Perimeter:_____.

Area _____.

4)

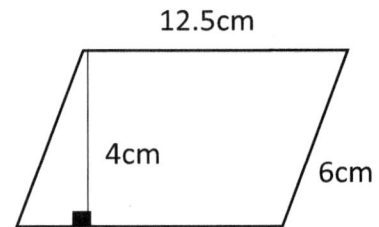

Perimeter:_____.

Area:_____.

5)

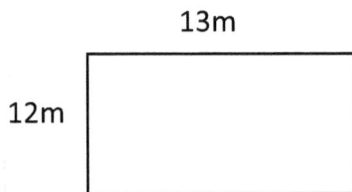

Perimeter:_____.

Area:_____.

6)

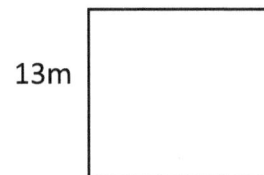

Perimeter:_____.

Area:_____.

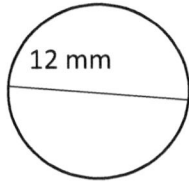

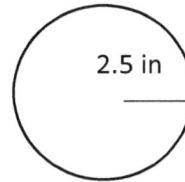

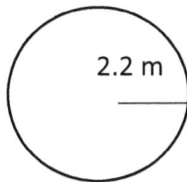

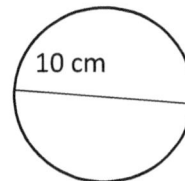

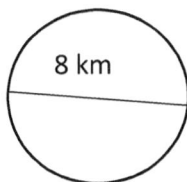

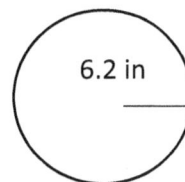

Circumference and Area of Circle

Find the circumference and area of each ($\pi = 3.14$).

1)

12 mm

Circumference:

Area:

2)

2.5 in

Circumference:_____.

Area:_____.

3)

2.2 m

Circumference:_____.

Area _____.

4)

10 cm

Circumference:_____.

Area:_____.

5)

8 km

Circumference:_____.

Area:_____.

6)

6.2 in

Circumference:_____.

Area:_____.

Perimeter of Polygon

Find the perimeter of each polygon.

1)

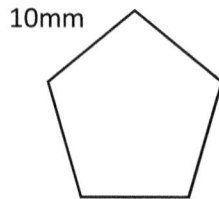

10mm

Perimeter:_____.

2)

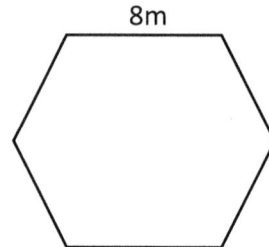

8m

Perimeter:_____:

3)

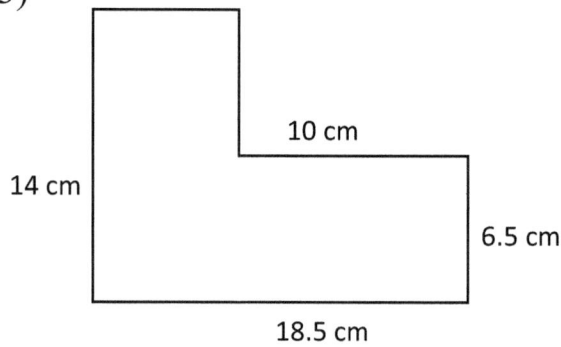

10 cm

14 cm

6.5 cm

18.5 cm

Perimeter:_____.

4)

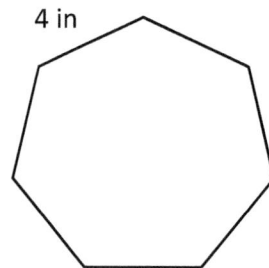

4 in

Perimeter:_____:

5)

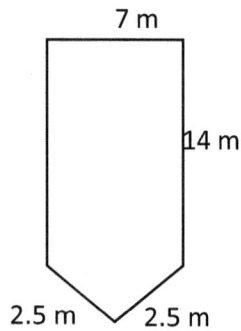

7 m

14 m

2.5 m   2.5 m

Perimeter:_____.

6)

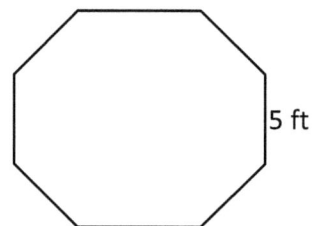

5 ft

Perimeter:_____:

Volume of Cubes

Find the volume of each cube.

1)

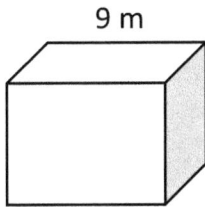

9 m

V:................................

2)

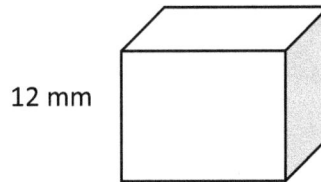

12 mm

V:................................

3)

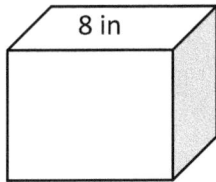

8 in

V:................................

4)

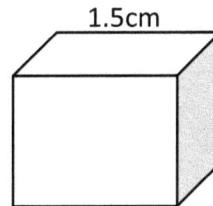

1.5cm

V:................................

5)

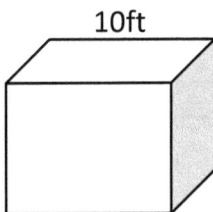

10ft

V:................................

6)

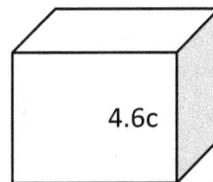

4.6c

V:................................

Volume of Rectangle Prism

Find the volume of each rectangle prism.

1)

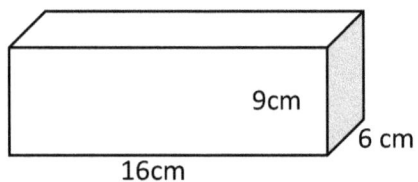

V:_____.

2)

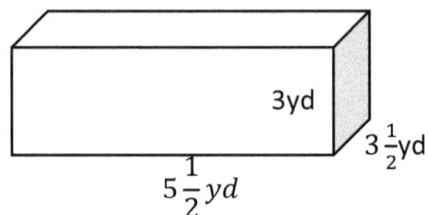

V:_____.

3)

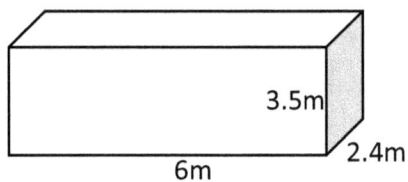

V:_____.

4)

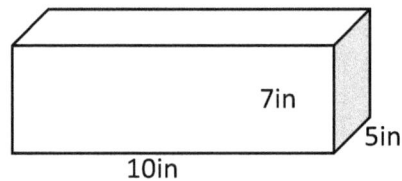

V:_____.

5)

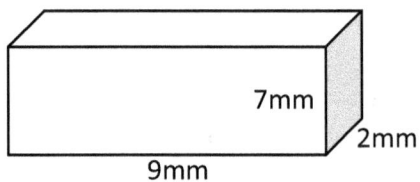

V:_____.

6)

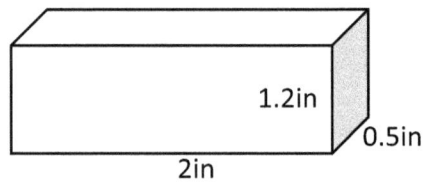

V:_____.

Volume of Cylinder

Find the volume of each cylinder.

1)

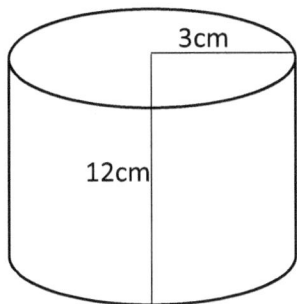

3cm

12cm

V:_____:

2)

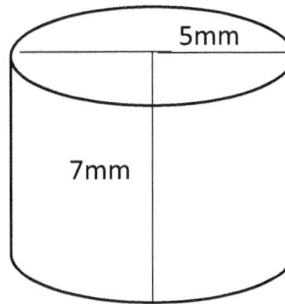

5mm

7mm

V:_____:

3)

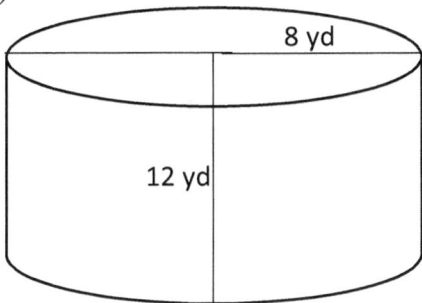

8 yd

12 yd

V:_____:

4)

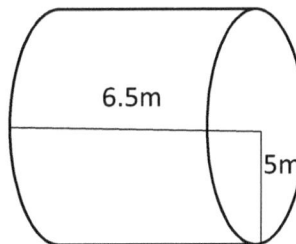

6.5m

5m

V:_____:

5)

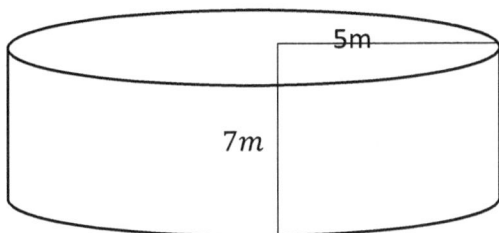

5m

7m

V:_____:

6)

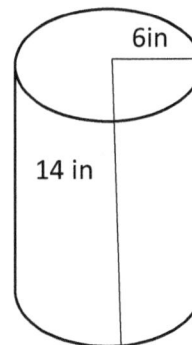

6in

14 in

V:_____:

Volume of Spheres

Find the volume of each spheres ($\pi = 3.14$).

1)

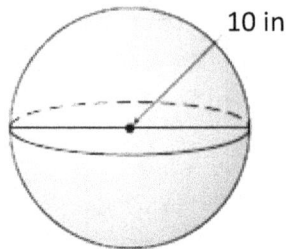

10 in

V:...

2)

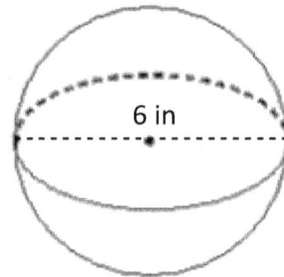

6 in

V:...

3)

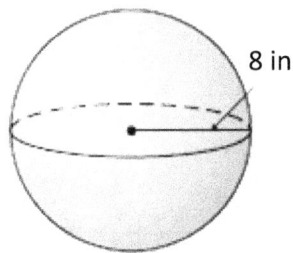

8 in

V:...

4)

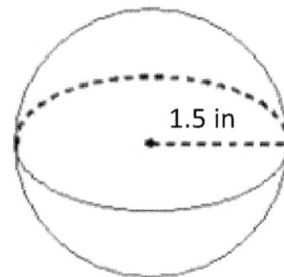

1.5 in

V:...

5)

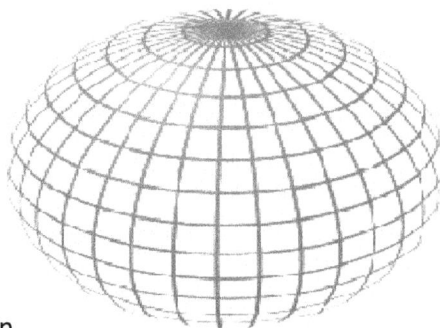

$r = 4\frac{1}{2}$ in

V:...

6)

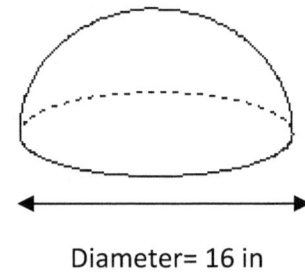

Diameter= 16 in

V:...

Volume of Pyramid and Cone

Find the volume of each pyramid and cone ($\pi = 3.14$).

1)

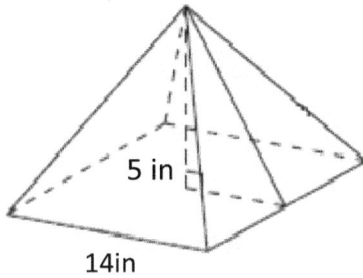

V:_____:

2)

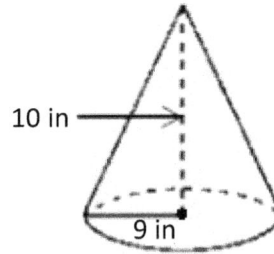

V:_____:

3)

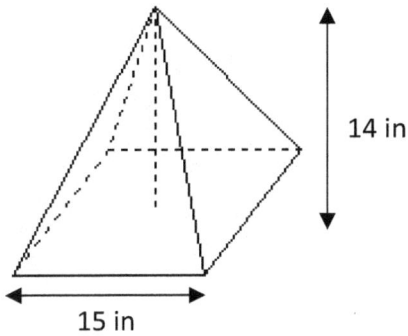

V:_____:

4)

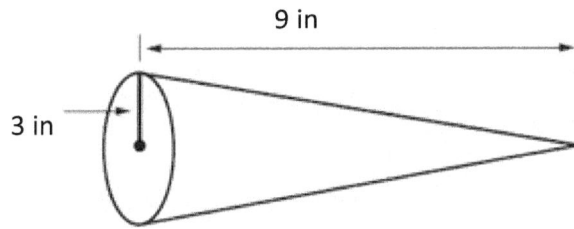

V:_____:

5)

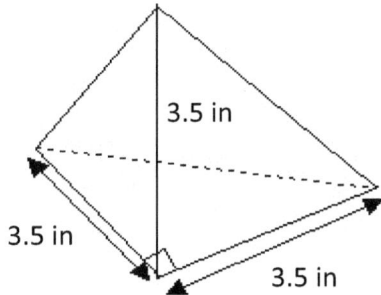

V:_____:

6)

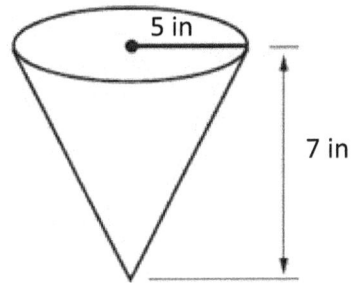

V:_____:

Surface Area Cubes

Find the surface area of each cube.

1)

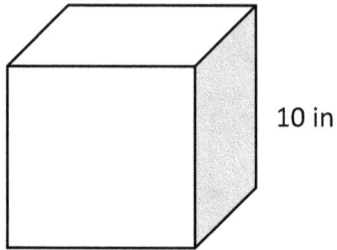

10 in

SA:_____:

2)

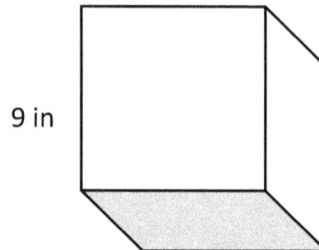

9 in

SA:_____:

3)

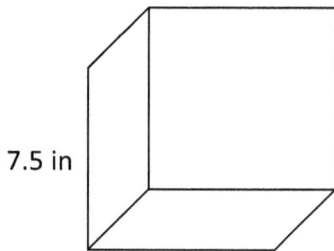

7.5 in

SA:_____:

4)

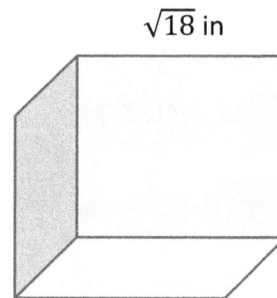

$\sqrt{18}$ in

SA:_____:

5)

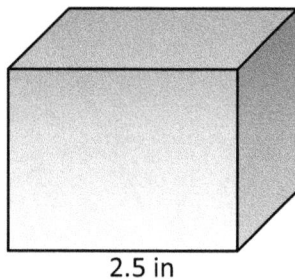

2.5 in

SA:_____:

6)

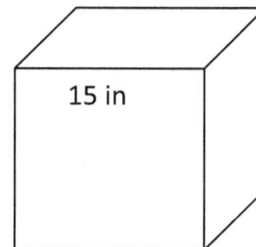

15 in

SA:_____:

Surface Area Rectangle Prism

Find the surface area of each rectangular prism.

1)

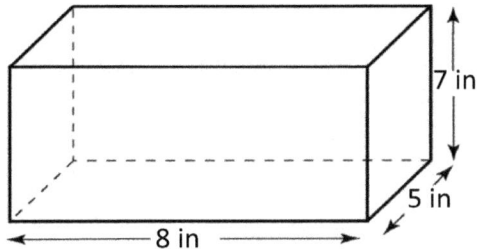

7 in

5 in

8 in

SA:_____:

2)

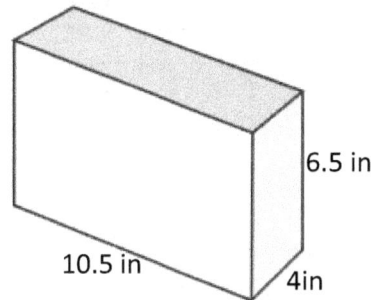

6.5 in

10.5 in

4 in

SA:_____:

3)

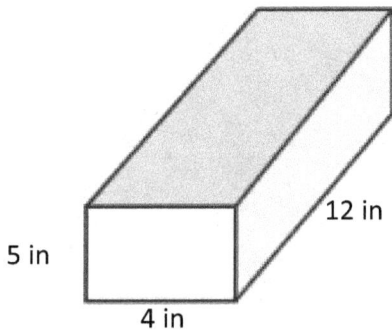

12 in

5 in

4 in

SA:_____:

4)

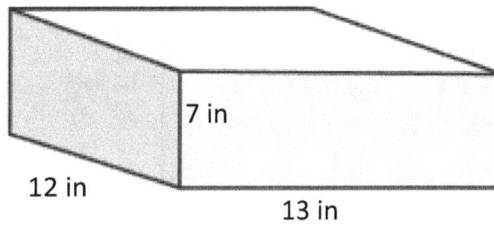

7 in

12 in

13 in

SA:_____:

5)

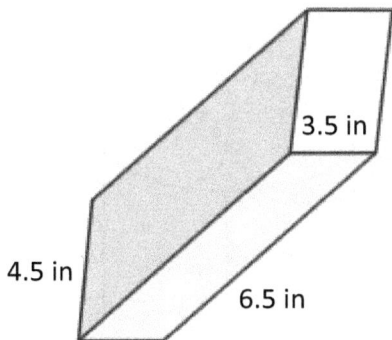

3.5 in

4.5 in

6.5 in

SA:_____:

6)

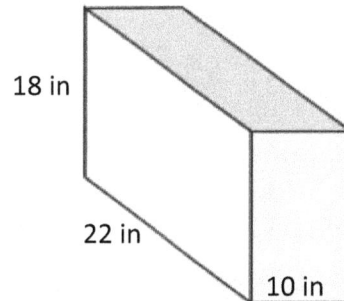

18 in

22 in

10 in

SA:_____:

Surface Area Cylinder

Find the surface area of each cylinder.

1)

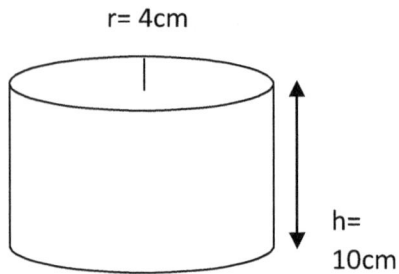

r= 4cm

h= 10cm

SA:_____.

2)

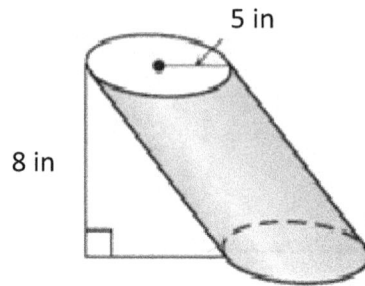

5 in

8 in

SA:_____.

3)

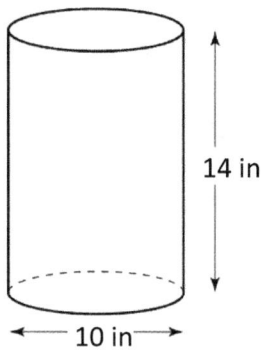

14 in

10 in

SA:_____.

4)

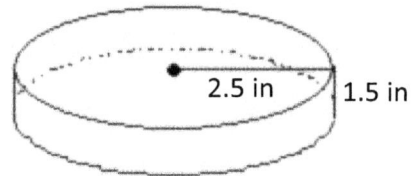

2.5 in   1.5 in

SA:_____.

5)

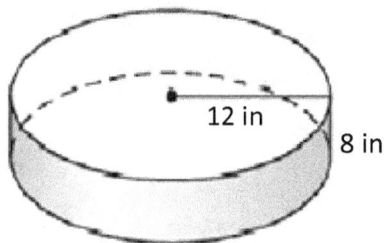

12 in

8 in

SA:_____.

6)

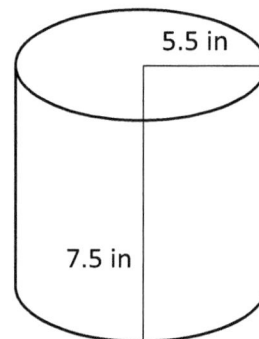

5.5 in

7.5 in

SA:_____.

Answers of Worksheets

Area and Perimeter of Square

1. Perimeter: 42, Area:49
2. Perimeter: $4\sqrt{7}$, Area:7
3. Perimeter: 32, Area:64
4. Perimeter: $4\sqrt{3}$, Area:3
5. Perimeter: 48, Area:144
6. Perimeter: $4\sqrt{32}$, Area:32

Area and Perimeter of Rectangle

1- Perimeter: 36, Area:56
2- Perimeter: 32, Area:63
3- Perimeter: 40, Area:91
4- Perimeter: 19, Area: 12
5- Perimeter: 12, Area: 8.64
6- Perimeter: 28, Area:48

Area and Perimeter of Triangle

1- Perimeter: 3s, Area:$\frac{1}{2}sh$
2- Perimeter: 60, Area:120
3- Perimeter: 32, Area:32
4- Perimeter: 36, Area:48
5- Perimeter: 24, Area:24
6- Perimeter: 96, Area:386

Area and Perimeter of Trapezoid

1- Perimeter: 80, Area:144
2- Perimeter: 15, Area:16
3- Perimeter: 41, Area:62
4- Perimeter: 38, Area:80
5- Perimeter: 44, Area:104
6- Perimeter: 52, Area:140

Area and Perimeter of Parallelogram

1- Perimeter: $26m$, Area:$30(m)^2$
2- Perimeter: $58m$, Area:$120(m)^2$
3- Perimeter: $44in$, Area:$60(in)^2$
4- Perimeter: $37cm$, Area:$50(cm)^2$
5- Perimeter: $50m$, Area:$156(m)^2$
6- Perimeter: $52m$, Area:$169(m)^2$

Circumference and Area of Circle

1) Circumference:37.68 mm   Area:$113.04(mm)^2$
2) Circumference: 15.7 in   Area:$19.625(in)^2$
3) Circumference: 13.816 m   Area:$15.197(m)^2$
4) Circumference: 31.4 cm   Area:$78.5(cm)^2$
5) Circumference: 25.12 in   Area:$50.24(in)^2$
6) Circumference: 38.936 km   Area:$120.702(km)^2$

Perimeter of Polygon

1) 50 mm
2) 48 m
3) 65 cm
4) 28 in
5) 40 m
6) 40 ft

Volume of Cubes

1) $729m^3$
2) $1,728(mm)^3$
3) $512in^3$
4) $3.375(cm)^3$

5) $1,000(ft)^3$ 6) $97.336(cm)^3$

Volume of Rectangle Prism

1) $864(cm)^3$ 3) $50.4(m)^3$ 5) $126(mm)^3$

2) $57.75(yd)^3$ 4) $350(in)^3$ 6) $1.2(in)^3$

Volume of Cylinder

1) $339.12(cm)^3$ 3) $602.88(yd)^3$ 5) $549.5(m)^3$

2) $549.5(mm)^3$ 4) $510.25(m)^3$ 6) $1,582.56(in)^3$

Volume of Spheres

1) $523.33(in)^3$ 3) $2,143.57(in)^3$ 5) $381.51(in)^3$

2) $113.04(in)^3$ 4) $14.13(in)^3$ 6) $1,071.79(in)^3$

Volume of Pyramid and Cone

1) $326.67\ (in)^3$ 3) $1,050\ (in)^3$ 5) $7.15\ (in)^3$

2) $847.8\ (in)^3$ 4) $84.78(in)^3$ 6) $183.17(in)^3$

Surface Area Cubes

1) $600(in)^2$ 3) $337.5(in)^2$ 5) $37.5(in)^2$

2) $486(in)^2$ 4) $108(in)^2$ 6) $1,350(in)^2$

Surface Area Rectangle Prism

1) $262(in)^2$ 3) $256(in)^2$ 5) $135.5(in)^2$

2) $272.5(in)^2$ 4) $662(in)^2$ 6) $1,592(in)^2$

Surface Area Cylinder

1) $351.68(in)^2$ 3) $596.6(in)^2$ 5) $1,507.2(in)^2$

2) $408.2(in)^2$ 4) $62.8(in)^2$ 6) $449.02(in)^2$

Chapter 10: Statistics and probability

Mean, Median, Mode, and Range of the Given Data

Find the mean, median, mode(s), and range of the following data.

1) 10, 2, 38, 23, 47, 23, 21

Mean: __, Median: __, Mode: __, Range: __

2) 12, 26, 26, 38, 30, 20

Mean: __, Median: __, Mode: __, Range: __

3) 41, 24, 49, 11, 45, 27, 35, 19, 24

Mean: __, Median: __, Mode: __, Range: __

4) 25, 11, 1, 15, 25, 18

Mean: __, Median: __, Mode: __, Range: __

5) 24, 14, 14, 17, 23, 15, 14, 29, 29, 8

Mean: __, Median: __, Mode: __, Range: __

6) 7, 14, 19, 11, 8, 19, 8, 15

Mean: __, Median: __, Mode: __, Range: __

7) 29, 28, 66, 76, 14, 44, 18, 44, 22, 44

Mean: __, Median: __, Mode: __, Range: __

8) 35, 35, 57, 78, 59

Mean: __, Median: __, Mode: __, Range: __

9) 16, 16, 29, 46, 54

Mean: __, Median: __, Mode: __, Range: __

10) 13, 9, 3, 3, 5, 6, 7

Mean: __, Median: __, Mode: __, Range: __

11) 4, 12, 4, 6, 1, 8

Mean: __, Median: __, Mode: __, Range: __

12) 8, 9, 15, 15, 17, 17, 17

Mean: __, Median: __, Mode: __, Range: __

13) 7, 7, 1, 16, 1, 7, 19

Mean: __, Median: __, Mode: __, Range: __

14) 13, 17, 10, 12, 12, 18, 15, 19

Mean: __, Median: __, Mode: __, Range: __

15) 29, 14, 30, 19, 29

Mean: __, Median: __, Mode: __, Range: __

16) 6, 6, 16, 18, 15, 22, 37

Mean: __, Median: __, Mode: __, Range: __

17) 25, 11, 14, 25, 18, 13, 7, 5

Mean: __, Median: __, Mode: __, Range: __

18) 55, 34, 34, 48, 85, 7

Mean: __, Median: __, Mode: __, Range: __

19) 54, 28, 28, 65, 5, 8

Mean: __, Median: __, Mode: __, Range: __

20) 77, 94, 25, 24, 11, 77, 19

Mean: __, Median: __, Mode: __, Range: __

Box and Whisker Plot

1) Draw a box and whisker plot for the data set:

   16, 11, 14, 12, 14, 12, 16, 16, 20

2) The box-and-whisker plot below represents the math test scores of 20 students.

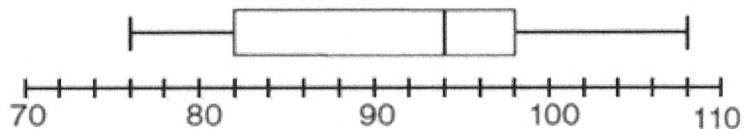

   A. What percentage of the test scores are less than 82?

   B. Which interval contains exactly 50% of the grades?

   C. What is the range of the data?

   D. What do the scores 76, 94, and 108 represent?

   E. What is the value of the lower and the upper quartile?

   F. What is the median score?

Bar Graph

Each student in class selected two games that they would like to play. Graph the given information as a bar graph and answer the questions below:

Game	Votes
Football	13
Volleyball	10
Basketball	18
Baseball	17
Tennis	13

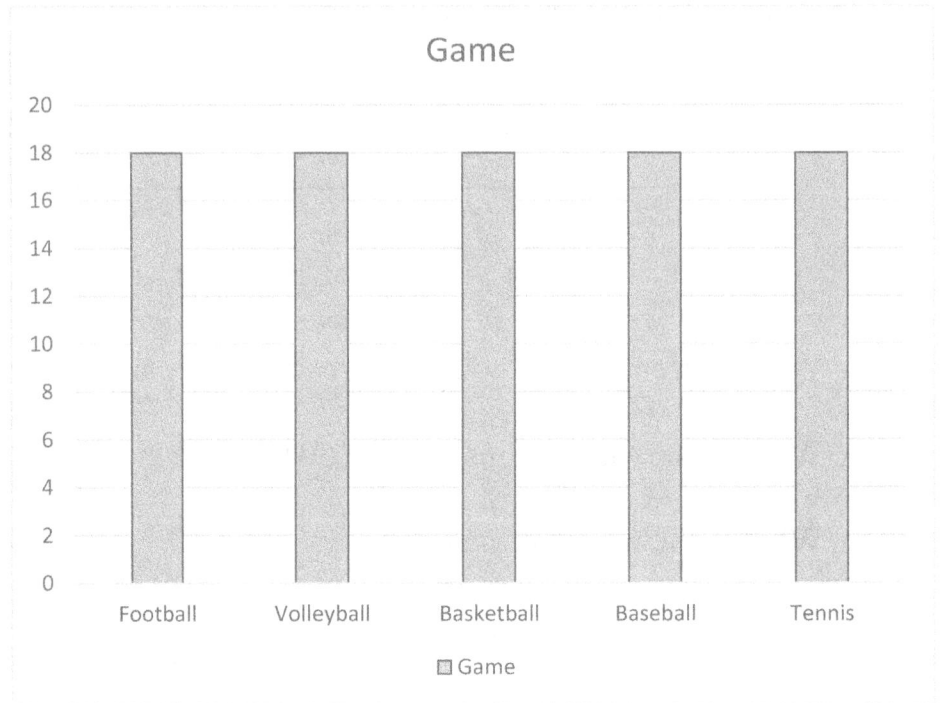

1) Which was the most popular game to play?

2) How many more student like Basketball than Volleyball?

3) Which two game got the same number of votes?

4) How many Volleyball and Football did student vote in all?

5) Did more student like football or Volleyball?

6) Which game did the fewest student like?

Histogram

Create a histogram for the set of data.

Math Test Score out of 100 points.

58	74	63	80	83	65	70	86	67	54
81	73	82	75	71	56	87	66	74	72
84	55	76	73	67	85	69	68	52	87

Frequency Table	
Interval	**Number of Values**

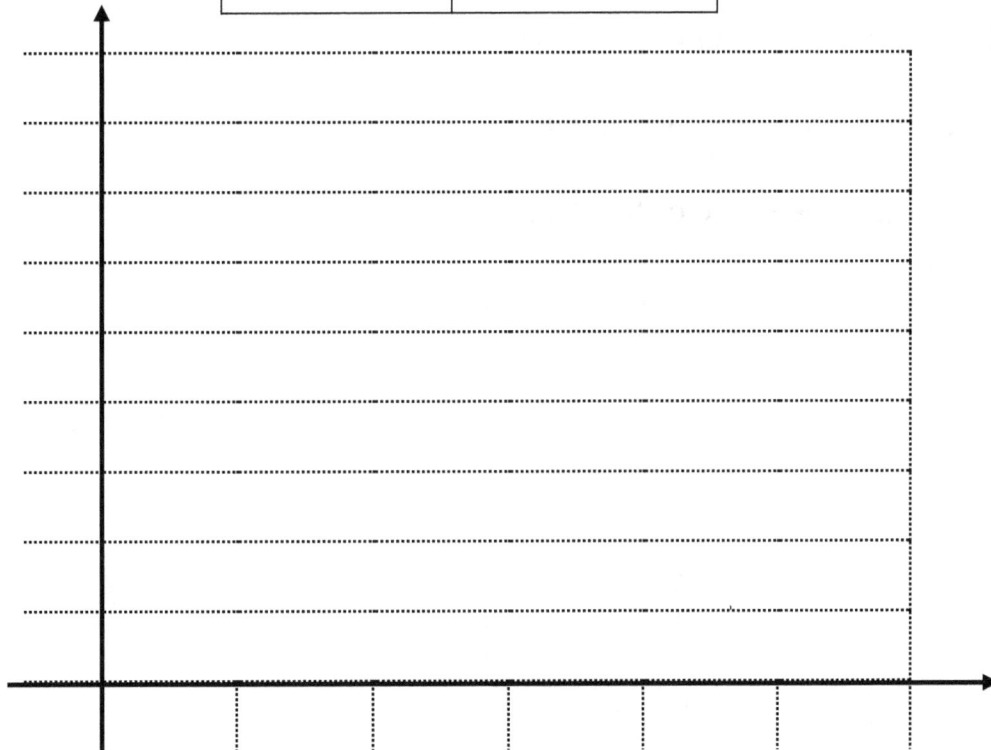

Dot plots

The ages of students in a Math class are given below.

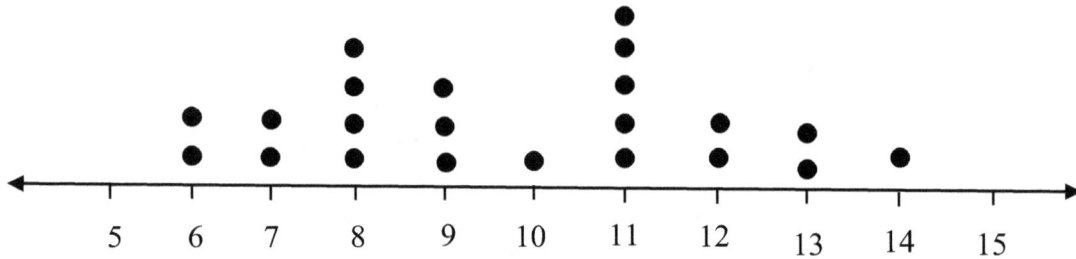

1) What is the total number of students in math class?

2) How many students are at least 12 years old?

3) Which age(s) has the most students?

4) Which age(s) has the fewest student?

5) Determine the median of the data.

6) Determine the range of the data.

7) Determine the mode of the data.

Scatter Plots

A person charges an hourly rate for his services based on the number of hours a job takes.

Hours	Rate
1	$24.5
2	$22
3	$21
4	$19.50

Hours	Rate
5	$19
6	$17.50
7	$17
8	$16.5

1) Draw a scatter plot for this data.

2) Does the data have positive or negative correlation?

3) Sketch the line that best fits the data.

4) Find the slope of the line.

5) Write the equation of the line using slope-intercept form.

6) Using your prediction equation: If a job takes 10 hours, what would be the hourly rate?

Stem–And–Leaf Plot

Make stem-and-leaf plots for the given data.

1) 17, 16, 58, 51, 12, 64, 18, 57, 59, 54, 19, 52, 65

Stem	leaf

2) 72, 74, 19, 41, 72, 12, 66, 78, 68, 64, 69, 62

Stem	leaf

3) 125, 108, 65, 65, 105, 127, 62, 126, 68, 124, 66, 109

Stem	leaf

4) 61, 45, 66, 60, 99, 63, 90, 97, 68, 63, 49, 42

Stem	leaf

5) 55, 58, 105, 56, 15, 108, 102

Stem	leaf

6) 143, 37, 87, 35, 140, 147, 83, 144, 38, 143, 89, 81

Stem	leaf

Pie Graph

60 people were survey on their favorite ice cream. The pie graph is made according to their responses. Answer following questions based on the Pie graph.

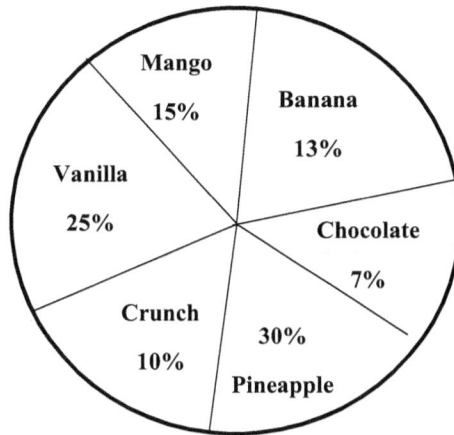

1) How many people like to eat Mango ice cream? _____

2) Approximately, which two ice creams did about half the people like the best? _____

3) How many people said either mango or crunch ice cream was their favorite? _____

4) How many people would like to have crunch ice cream? _____

5) Which ice cream is the favorite choice of 15 people? _____

Probability

1) A jar contains 16 caramels, 5 mints and 19 dark chocolates. What is the probability of selecting a mint?

2) If you were to roll the dice one time what is the probability it will NOT land on a 4?

3) A die has sides are numbered 1 to 6. If the cube is thrown once, what is the probability of rolling a 5?

4) The sides of number cube have the numbers 4, 6, 8, 4, 6, and 8. If the cube is thrown once, what is the probability of rolling a 6?

5) Your friend asks you to think of a number from ten to twenty. What is the probability that his number will be 15?

6) A person has 8 coins in their pocket. 2 dime, 3 pennies, 2 quarter, and a nickel. If a person randomly picks one coin out of their pocket. What would the probability be that they get a penny?

7) What is the probability of drawing an odd numbered card from a standard deck of shuffled cards (Ace is one)?

8) 32 students apply to go on a school trip. Three students are selected at random. what is the probability of selecting 4 students?

Answers of Worksheets

Mean, Median, Mode, and Range of the Given Data

1) mean: 23.43, median: 23, mode: 23, range: 45.

2) mean: 25.33, median: 26, mode: 26, range: 26.

3) mean: 30.56, median: 27, mode: 24, range: 38.

4) mean: 15.83, median: 16.5, mode: 25, range: 24.

5) mean: 18.7, median: 16, mode: 14, range: 21.

6) mean: 12.63, median: 12.5, mode: 19, 8, range: 12.

7) mean: 38.5, median: 36.5, mode: 44, range: 62.

8) mean: 52.8, median: 57, mode: 35, range: 43.

9) mean: 32.2, median: 29, mode: 16, range: 38.

10) mean: 6.57, median: 6, mode: 3, range: 10.

11) mean: 5.83, median: 5, mode: 4, range: 11.

12) mean: 14, median: 15, mode: 17, range: 9.

13) mean: 8.29, median: 7, mode: 7, range: 18.

14) mean: 14.5, median: 14, mode: 12, range: 9.

15) mean: 24.2, median: 29, mode: 29, range: 16.

16) mean: 17.14, median: 16, mode: 6, range: 31.

17) mean: 14.75, median: 13.5, mode: 25, range: 20.

18) mean: 43.83, median: 41, mode: 34, range: 78.

19) mean: 31.33, median: 28, mode: 28, range: 60.

20) mean: 46.71, median: 25, mode: 77, range: 83.

Box and Whisker Plot

1)

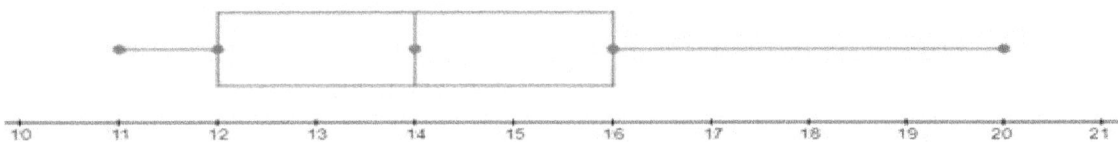

2)

A. 25%

B. 94

C. 32

D. Minimum, Median, and Maximum.

E. Lower (Q_1) is 82 and upper (Q_3)is 98.   F. 94

Bar Graph

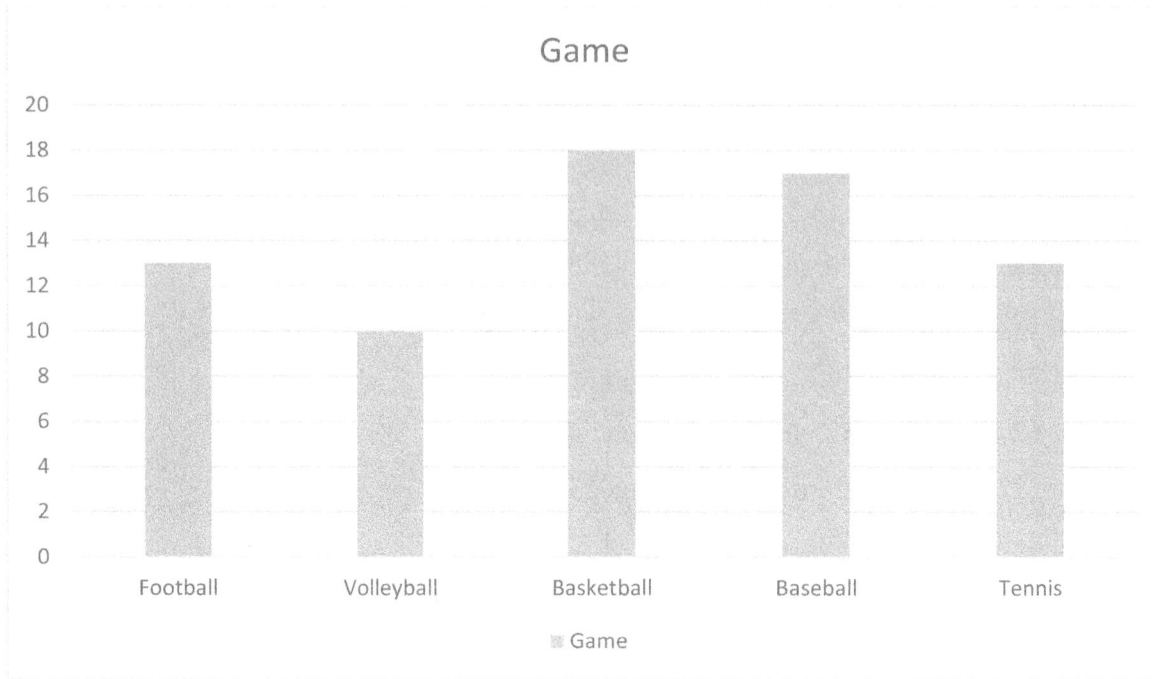

1) Basketball   3) Football and Tennis   5) Football

2) 8 students   4) 23   6) Volleyball

Histogram

Frequency Table	
Interval	**Number of Values**
52-57	4
58-63	2
64-69	6
70-75	8
76-81	3
82-87	7

Histogram

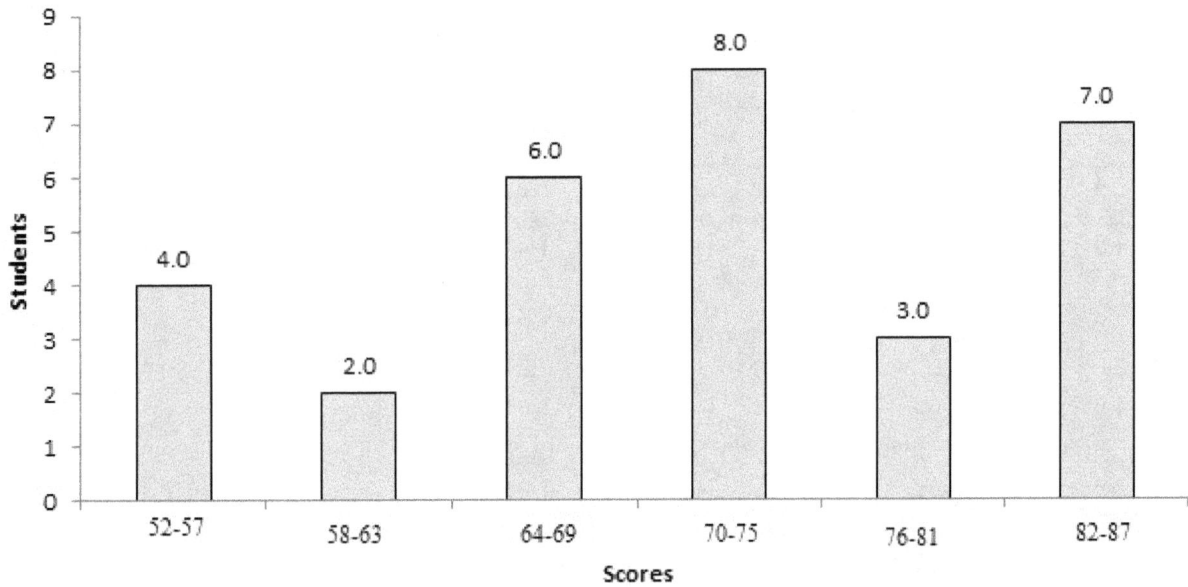

Dot plots

1) 22

2) 5

3) 11

4) 10 and 14

5) 2

6) 4

7) 2

Scatter Plots

1)

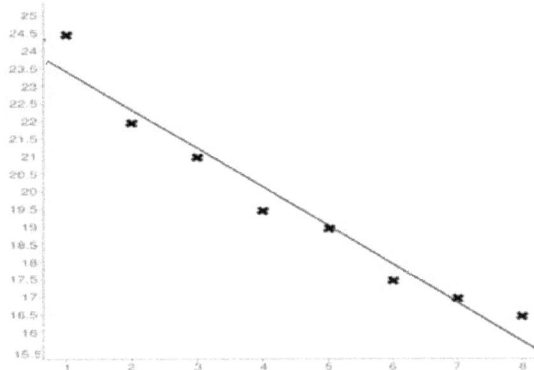

2) Negative correlation

3) ----

4) Slope(m)= -1

5) $y = -x + 24.5$

6) 14.5

Stem–And–Leaf Plot

1)

Stem	leaf
1	2 6 7 8 9
5	1 2 4 7 8 9
6	4 5

2)

Stem	leaf
1	2 9
6	1 2 4 6 8 9
7	2 2 4 8

3)

Stem	leaf
6	2 5 5 6 8
10	5 8 9
12	4 5 6 7

4)

Stem	leaf
4	2 9 5
6	0 1 3 3 6 8
9	0 7 9

5)

Stem	leaf
1	5
5	5 6 8
10	2 5 8

6)

Stem	leaf
3	5 7 8
8	1 3 7 9
14	0 3 3 4 7

Pie Graph

1) 9

2) Vanilla and pineapple

3) 15

4) 6

5) Vanilla

Probability

1) $\frac{1}{8}$

2) $\frac{5}{6}$

3) $\frac{1}{6}$

4) $\frac{1}{3}$

5) $\frac{1}{10}$

6) $\frac{3}{8}$

7) $\frac{5}{13}$

8) $\frac{1}{8}$

GED Mathematics

Test Review

GED Mathematics Test Formula Sheet

Area of a:	
Parallelogram	$A = bh$
Trapezoid	$A = \dfrac{1}{2}h(b_1 + b_2)$

Surface Area and Volume of a:		
Rectangular/Right Prism	$SA = ph + 2B$	$V = Bh$
Cylinder	$SA = 2\pi rh + 2\pi r^2$	$V = \pi r^2 h$
Pyramid	$SA = \dfrac{1}{2}ps + B$	$V = \dfrac{1}{3}Bh$
Cone	$SA = \pi rs + \pi r^2$	$V = \dfrac{1}{3}\pi r^2 h$
Sphere	$SA = 4\pi r^2$	$V = \dfrac{4}{3}\pi r^3$
	(p = perimeter of base B; π = 3.14)	

Algebra	
Slope of a line	$m = \dfrac{y_2 - y_1}{x_2 - x_1}$
Slope-intercept form of the equation of a line	$y = mx + b$
Point-slope form of the Equation of a line	$y - y_1 = m(x - x_1)$
Standard form of a Quadratic equation	$y = ax^2 + bx + c$
Quadratic formula	$x = \dfrac{-b \pm \sqrt{b^2 - 4ac}}{2a}$
Pythagorean theorem	$a^2 + b^2 = c^2$
Simple interest	$I = prt$ (I = interest, p = principal, r = rate, t = time)

GED Practice Test 1

Mathematical Reasoning

Two Parts

Total Number of Questions: 46

Part 1 (Non-Calculator): 5 Questions

Part 2 (Calculator): 41 Questions

Total time for two Part: 115 Minutes

Administered *Month Year*

GED Practice Test 1
Mathematical Reasoning
Part 1: Non-Calculator
You may Not use a calculator in this part.

1) How many odd integers are between $\frac{-29}{4}$ and $\frac{23}{5}$?

    A. 4

    B. 7

    C. 6

    D. 9

2) Which expression correctly represents the distance between the two points shown on the number line?

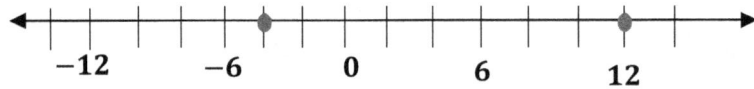

    A. $-4 - 12$

    B. $|-12 + 4|$

    C. $-4 + 12$

    D. $|12 + 4|$

3) In the triangle ABC, if angle B and angle C both equal 60°, then what is the

length of side BC?

A. 6 cm

B. 12 cm

C. 18 cm

D. 24 cm

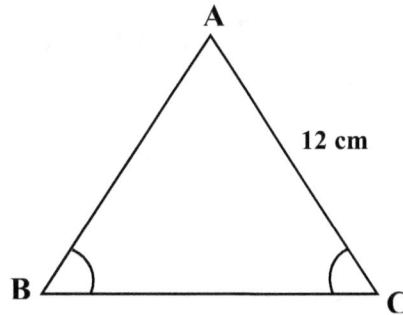

4) Which of the following represents the sum of the factors of 20?

A. 42

B. 47

C. 41

D. 39

5) Alice read $\frac{1}{3}$ of her book on Saturday and $\frac{3}{5}$ of her book on Sunday. What

fraction of her book remains to be read?

A. $\frac{1}{20}$

B. $\frac{1}{15}$

C. $\frac{14}{15}$

D. $\frac{3}{20}$

GED Practice Test 1

Mathematical Reasoning

Part 2: Calculator

You may use a calculator in this part.

6) Consider the following series: 2, 4, 3, 5, 9, 12. What number should come next?

    A. 21

    B. 27

    C. 17

    D. 16

7) Find the greatest common factor of 12, 21, and 32.

    A. 1

    B. 3

    C. 4

    D. 12

8) Which equation can be equal "5 more than the ratio of a number to 7 is equal to 4 less than the number"?

    A. $5x - 7 = 4 - x$

    B. $5 + \frac{x}{7} = x - 4$

    C. $\frac{5}{7}x - 4 = 7x$

    D. $5 + 7x = 4 - x$

9) Arrange the following fractions in order from least to greatest.

$$\frac{4}{9}, \frac{7}{12}, \frac{1}{5}, \frac{14}{15}, \frac{16}{21}$$

A. $\frac{1}{5}, \frac{4}{9}, \frac{7}{12}, \frac{16}{21}, \frac{14}{15}$

B. $\frac{7}{12}, \frac{4}{9}, \frac{1}{5}, \frac{14}{15}, \frac{16}{21}$

C. $\frac{16}{21}, \frac{14}{15}, \frac{7}{12}, \frac{4}{9}, \frac{1}{5}$

D. $\frac{14}{15}, \frac{16}{21}, \frac{1}{5}, \frac{4}{9}, \frac{7}{12}$

10) Elena earns \$7.60 an hour and worked 35 hours. Her brother earns \$13.30 an hour. How many hours would her brother need to work to equal Elena's earnings over 25 hours?

A. 24.2

B. 20

C. 30

D. 42.50

11) $31.84 \div 0.8 =$

A. 3.98

B. 39.80

C. 33.98

D. 3.398

12) What is the circumference of a circle with a radius of 14 inches?

   A. $144\,\pi$

   B. $28\,\pi$

   C. $36\,\pi$

   D. $48\,\pi$

13) Alfred needs to calculate his monthly water bill. His family used 26,200 gallons at a rate of $0.65 per hundred gallons. Also, there is a monthly fee of $4.80 on each period. What is his total bill?

   A. $15,710

   B. $157.1

   C. $175.1

   D. $275.1

14) Simplify $\dfrac{(4x^5 - 8x^2)}{(2x^4 - 4x)} = ?$

   A. $2x$

   B. $2x - 1$

   C. $4x$

   D. $2x(x - 1)$

15) Which of the following expressions is undefined in the set of real numbers?

   A. $\sqrt[2]{169}$

   B. $\sqrt[3]{-27}$

   C. $\sqrt{-81}$

   D. $\sqrt[4]{16}$

16) If $f(x) = 3x^2$, and 2f(4a) = 864 then what could be the value of a?

   A. -2

   B. -1

   C. 5

   D. 3

17) Speed of a train is 108 mile per hour for 2 hours and 25 minutes. How many miles did the train travel?

   A. 243

   B. 279

   C. 227

   D. 234

18) Solve $14x - 8 \le 16x + 4$

   A. $x \le -6$

   B. $x \ge -6$

   C. $x \le 6$

   D. $x \ge 6$

19) What is the value of 3^8?

    A. $(3 + 3)^6$

    B. 9^3

    C. $3(3^2)$

    D. $3^4 + 3^4$

20) Shane Williams puts $3,800 into a saving bank account that pays simple interest of 2.75%. How much interest will she earn after 4 years?

    A. $4,448

    B. $ 4,148

    C. $418

    D. $841

21) Three angles join to form a straight angle. One angle measure 70°. Other angle measures 38°. What is the measure of third angle?

    A. 28°

    B. 108°

    C. 58°

    D. 72°

22) Evaluate $\dfrac{56x^5y^4z^{-4}}{16x^2\,y^6z^3}$.

    A. $\dfrac{7x^3y^2}{2\,z^7}$

    B. $\dfrac{3x^3z^2}{2\,y^3}$

    C. $\dfrac{7x^3}{2\,y^2z^7}$

    D. $\dfrac{7y^3}{2\,x^2z^2}$

23) Find the equation for line passing through $(4,-2)$ and $(7,2)$.

    A. $-2x - 3y = 24$

    B. $3y - 4x = -22$

    C. $-4x + 2y = 20$

    D. $4y + 3x = 22$

24) If $x = -2$ and $y = 4$, calculate the value of $\dfrac{x^3-2}{y+1}$.

    A. -1

    B. -3

    C. -2

    D. 1

25) The table shows the parking rates for the outside terminal area at an airport.

Ethan parked at the lot for $1\frac{3}{4}$ hours. How much did he owe?

A. $4.75

B. $3.80

C. $5.75

D. $5.25

1 hours	$3.8
Each 15 minutes after 1 hours	$0.65
24-hours Discount rate	$2.8

26) What is the value of x in term of c and d $\frac{c+d}{3dx} = \frac{2}{7}$? (c and d>0)

A. $\frac{7}{6}(\frac{c}{d} + 1)$

B. $\frac{7}{6}(1 - \frac{c}{d})$

C. $\frac{6}{7}\left(\frac{c}{d} + 1\right)$

D. $\frac{6}{7}(1 - \frac{c}{d})$

27) Factor the equation $x^4 - 9x^2$.

A. $9x(x - 1)$

B. $3x(x + 3)(x - 3)$

C. $x^2(x - 5)(x + 4)$

D. $x^2(x - 3)(x + 3)$

28) Which of the following are the solutions to the equation $x^2 - 13x + 42 = 0$?

    A. $-13, 6$

    B. $6, 7$

    C. $-6, -7$

    D. $13, -7$

29) Which of the following equations best represents the line in the graph below?

    A. $y = \frac{1}{4}x + 3$

    B. $y = 4x + 6$

    C. $y = \frac{1}{4}x - 3$

    D. $y = 4x + 6$

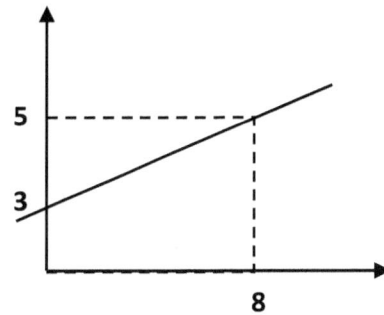

30) If $-2x + 4y = -6$ and $x - 3y = 5$, what is the value of x?

    A. 2

    B. 1.4

    C. -3

    D. -1

31) lengths of Two sides of a triangle are 3 and 8. Which of the following could Not be the measure of third side?

    A. 6

    B. 4

    C. 12

    D. 14

32) The following data set is given: 113, 101, 132, 153, 128, 159

Adding which number to the set will increase its mean?

A. 121

B. 129

C. 130

D. 141

33) An award for best education improvement is awarded annually to a winning

US state, and the winners from 2002 to 2011 are given in the table below. Find

the mode of this set of states.

Year	2002	2003	2004	2005	2006	2007	2008	2009	2010	2011
State	New Jersey	Ohio	New York	New York	Ohio	California	Ohio	Oregon	New York	New Jersey

A. New York and New Jersey

B. New York and Ohio

C. New Ohio and New York

D. Oregon and California

34) How long is a distance of 9 km if measured on a map with a scale of

1:30,000?

A. 34

B. 30

C. 13

D. 60

35) What is the area of a square if its side measures $\sqrt{5}$ m?

    A. $\sqrt{5}$

    B. $5\sqrt{5}$

    C. $2\sqrt{5}$

    D. 5

36) What is the answer of the following equation $3x^3y^2\,(2x^3y^4)^4 = ?$

    A. $24x^{12}y^{16}$

    B. $48x^{15}y^{18}$

    C. $48x^{12}y^{16}$

    D. $24x^{15}y^{18}$

37) The circle graph below shows the type of pizza that people prefer for lunch. If 150 people were surveyed, how many people preferred Peperoni?

    A. 35

    B. 25

    C. 30

    D. 20

Pizza

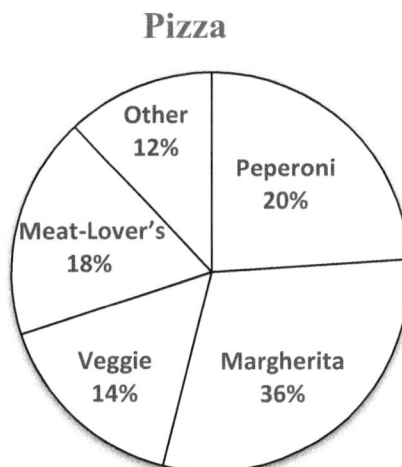

Circle graph: Other 12%, Peperoni 20%, Meat-Lover's 18%, Veggie 14%, Margherita 36%

38) A baseball has a volume of 36π. What is the length of the diameter?

   A. 3

   B. 6

   C. 9

   D. 27

39) 154 is What percent of 140?

   A. 110 %

   B. 60 %

   C. 10 %

   D. 140 %

40) A phone manufacturer makes 24,000 phone a year. The company randomly

   selects 600 of the phones to sample for inspection. The company discovers

   that there are 4 faulty phones in the sample. Based on the sample, how many

   of the 24,000 total phones are likely to be faulty?

   A. 240

   B. 100

   C. 120

   D. 160

41) Emma and Mia buy a total of 16 books. Emma bought 4 more books than Mia did. How many books did Emma buy?

    A. 12

    B. 10

    C. 2

    D. 20

42) A bag contains 7 white balls, 5 red balls and 13 black balls. A ball is picked from the bag at random. Find the probability of picking a red ball?

    A. 1.20

    B. 0.10

    C. 1.10

    D. 0.20

43) The radius of the following cylinder is 3 inches, and its height are 9 inches. What is the surface area of the cylinder in square inches? ($\pi=3.14$)

    A. 226.08

    B. 162.18

    C. 2,260

    D. 1,680

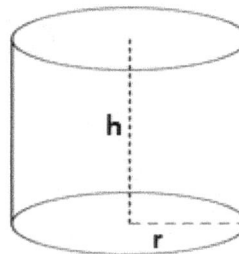

44) Find the length of the unknown side.

A. 13 ft

B. 26 ft

C. 28 ft

D. 36 ft

10 ft *24 ft* *? ft*

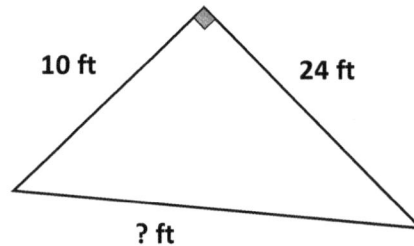

45) The line n has a slope of $\frac{a}{b}$, where c and d are integers. What is the slope of a line that is perpendicular to line n?

A. $\frac{a}{b}$

B. $-\frac{a}{b}$

C. $\frac{b}{a}$

D. $-\frac{b}{a}$

46) In a store, 24% of customers are female. If the total number of customers is 750, then how many male customers are dealing with the store?

A. 157

B. 370

C. 570

D. 720

"End of GED Mathematical Reasoning Practice Test 1."

GED Practice Test 2

Mathematical Reasoning

Two Parts

Total Number of Questions: 46

Part 1 (Non-Calculator): 5 Questions

Part 2 (Calculator): 41 Questions

Total time for two Part: 115 Minutes

Administered *Month Year*

GED Practice Test 2
Mathematical Reasoning
Part 1: Non-Calculator
You may Not use a calculator in this part.

1) What is the sum of the smallest prime number and five times the largest negative even integer?

   A. -2

   B. 12

   C. -8

   D. -10

2) Sam's incomes and expenditures for the first season of the last year are given in the table below. In which month were her savings the highest?

   A. March

   B. February

   C. January

   D. All months were the same.

Month	Income	Cost
January	$5,145	$3,990
February	$5,620	$3,885
March	$5,380	$3,780

3) Solve these fractions and reduce to its simplest terms: $8\frac{5}{18} - 5\frac{1}{6} + 2\frac{2}{3} =$

A. $5\frac{7}{9}$

B. $-5\frac{7}{18}$

C. $\frac{5}{9}$

D. $1\frac{5}{18}$

4) Calculate, $3^4 + 3^3 + 3 =?$

A. 3^8

B. 81

C. 111

D. 99

5) Which decimal is equivalent to $\frac{180}{450}$?

A. 1.40

B. 0.36

C. 0.04

D. 0.4

GED Practice Test 2

Mathematical Reasoning

Part 2: Calculator

You may use a calculator in this part.

6) Find the solution set of the following equation: $|2x - 4| = 8$

   A. $\{4, -2\}$

   B. $\{2, -6\}$

   C. $\{6\}$

   D. $\{-2, 6\}$

7) What is the solution to the pair of equations below? $\begin{cases} 3x - 5y = 6 \\ 2x + y = 4 \end{cases}$

   A. $x = -2$ and $y = 0$

   B. $x = 0$ and $y = -2$

   C. $x = 2$ and $y = 0$

   D. $x = -2$ and $y = 2$

8) The train each 24 minutes passes an average 5 stations. At this rate, how many stations will it pass in two hours.

   A. 20

   B. 25

   C. 15

   D. 30

9) If Sofia buys a shirt marked down 15 percent from its $130 while Natalia buys the same shirt mark down only 13 percent, how much more does Natalia pay for the shirt?

    A. $0.26

    B. $1.20

    C. $2

    D. $2.60

10) Find the circumference of the circle in terms of π.

    A. 13.5π in.

    B. 273π in.

    C. 54π in.

    D. $1,458\pi$ in.

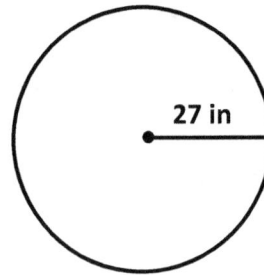

11) Find the volume of rectangular prism below?

    A. $562\ in^3$

    B. $156\ in^3$

    C. $1,056\ in^3$

    D. $1,760\ in^3$

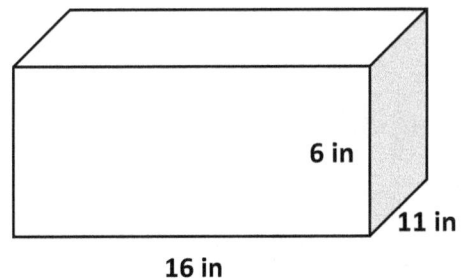

12) What is $\sqrt[5]{2^{-10}}$ in simplest form?

A. $\dfrac{1}{1,024}$

B. $\dfrac{1}{64}$

C. $\dfrac{1}{8}$

D. $\dfrac{1}{4}$

13) Which statement correctly describes the value of N in the equation below?

$7(6N - 32) = 6(7N - 12)?$

A. N has no correct solutions.

B. N=0 is one solution.

C. N has infinitely many correct solutions.

D. N=1 is one solution.

14) What is the maximum amount of grain, the silo can hold, in cubic feet?

A. $255\pi\ m^3$

B. $280\pi\ m^3$

C. $350\pi\ m^3$

D. $1,360\pi\ m^3$

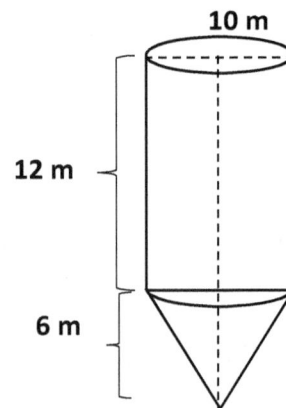

15) What is 9.23×10^{-6} in standard form?

   A.$- 923,000$

   B.$- 0.00000923$

   C.$\dfrac{1}{923,000}$

   D.0.00000923

16) What is the value of x in the triangle?

   A.$130°$

   B.$100°$

   C.$50°$

   D.$32°$

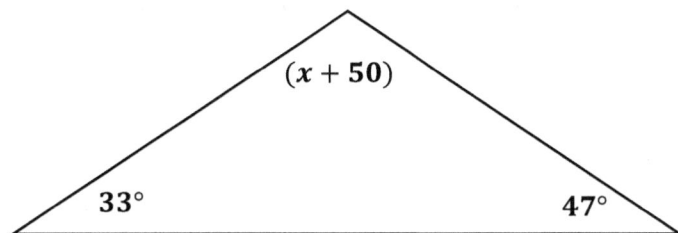

17) Find the length of the unknown side.

   A.12 ft

   B.8 ft

   C.10 ft

   D.5 ft

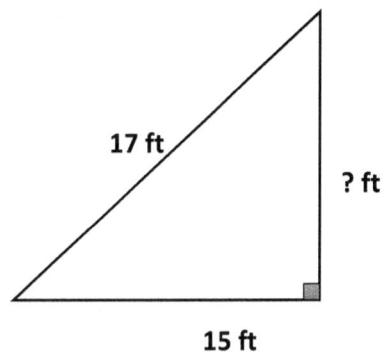

18) A store sells all of its products at a price 40% greater than the price the store paid for the product. How much does the store sell a product if the store paid $420 for it?

A.$252

B.$858

C.$588

D.$160

19) What is the area of shaded region?

A.132

B.48

C.180

D.100

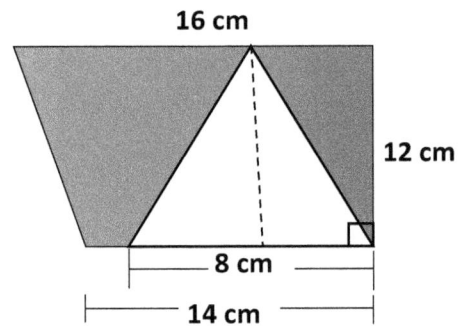

16 cm

12 cm

8 cm

14 cm

20) if $xy - 6x = 64$ and $y - 6 = 8$, then $x =$?

A.56

B.12

C.6

D.8

21) Which is the value of x in the equation $\frac{x}{3} = x - 8$?

    A. 18

    B. 6

    C. 12

    D. 24

22) What is the value of x, If $-3x - 2y = 5$ and $-4x - 3y = 7$?

    A. 0

    B. -3

    C. -4

    D. -1

23) What is the probability of Not spinning at H?

    A. $\frac{5}{8}$

    B. $\frac{3}{8}$

    C. $\frac{1}{2}$

    D. $\frac{1}{4}$

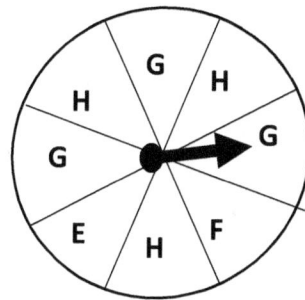

24) Ella bought 21 movies for \$2.06 per movie. Which equation shows the BEST estimate of the total cost?

    A. $20 \times \$2 = \$40$

    B. $20 \times \$3 = \$60$

    C. $21 \times \$2 = \$42$

    D. $21 \times \$3 = \$63$

25) A pick-up truck travels 60 mile on 12 L of gasoline when driven on a smooth road. If the cost of gasoline is $2.05/L, which is the cost of 1,500 mile of highway (smooth)?

 A. $16,115

 B. $1,516

 C. $156

 D. $615

26) A position of subway station and Grace's house shown by a grid. The station is located at $(-4, -5)$, and her house is located at $(8,4)$. What is the distance between her house and the subway stop?

 A. 12

 B. 15

 C. $15\sqrt{2}$

 D. 5

27) For the following set of numbers find the median.

 158, 47, 224, 83, 152, 98, 402.

 A. 152

 B. 224

 C. 156

 D. 402

28) What is solution to the equation $\sqrt{5x - 4} = 4$?

    A. -4

    B. -3

    C. 3

    D. 4

29) The equation $3x = 8y - 24$ has a y-intercept of?

    A. 3

    B. -24

    C. $\dfrac{3}{8}$

    D. $-\dfrac{3}{8}$

30) If $3^{2x} = 729$, then $x = ?$

    A. 2

    B. 4

    C. 3

    D. 6

31) Which is the smallest positive integer which is divisible by both 15 and 30.

    A. 5

    B. 15

    C. 30

    D. 450

32) A cube has total surface area of 24 cm². what is the volume of the cube in cm³?

   A. 9

   B. 8

   C. 6

   D. 3

33) Which is the value of x^2, if $x^2 + 5x = 36$?

   A. 4

   B. -16

   C. 36

   D. 81

34) Each of 9 pitchers can contain up to $\frac{4}{9}$ L of water. If each of the pitcher is at least the one-fourth full, which of the following expressions represents the total amount of water, W, contained on all 9 pitchers?

   A. $0.5 < w < 2.5$

   B. $2 < w < 2.5$

   C. $2 < w < 4$

   D. $1 < w < 4$

35) Ages of the players on a volleyball team is given. Which is the range of their

   ages? 39, 48, 51, 52, 37, 40, 47, 49, 36, 35, 50, 42

   A. 25

   B. 13

   C. 17

   D. 19

36) Find the area of the circle to the nearest tenth. Use 3.14 for π.

   A. 50.2

   B. 55.2

   C. 12.6

   D. 15.4

8 mm

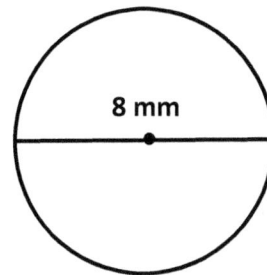

37) What is the simplest form of the expression$\frac{2x^2-11x-6}{4(x^2-\frac{1}{4})}$?

   A. $\frac{2x+5}{4(x-\frac{1}{4})}$

   B. $\frac{x-6}{2x-1}$

   C. $\frac{x+6}{2x+1}$

   D. $\frac{2x-1}{x-6}$

38) What is the value of $\left(\frac{1}{3}\right)^{-4}$?

A. $\frac{1}{81}$

B. -81

C. $-\frac{1}{81}$

D. 81

39) What is the simplest form of the expression $\frac{(2x^{-2}y^3)^4}{48y^{-5}z^{-4}}$, (using positive

exponent)?

A. $\frac{9y^{17}z^4}{x^8}$

B. $\frac{y^{17}z^4}{3x^8}$

C. $\frac{x^{12}}{9z^8y^{12}}$

D. $\frac{x^{17}}{3y^8z^3}$

40) A fruit sells for \$30 per kilograms. What is the price in cent per gram?

A. 0.003

B. 0.3

C. 0.03

D. 3

41) The price of water triples every 6 years. If the price of water on January 1st, 2012 is $9 per gallon, what is the equation that would be used to calculate the price(P) of water on January 1st, 2006?

A. $9P = 9$

B. $\dfrac{P}{3} = 9$

C. $3p = 9$

D. $9P = 3$

42) The line $4y - 5 = 12x + 11$ and $3y - 2 = x + 6$ are.

A. Perpendicular

B. Parallel

C. The same line

D. Neither parallel nor perpendicular

43) A rectangular box measures $4\frac{1}{2}$ feet by $3\frac{1}{5}$ feet. It is divided into four equal parts. What is the area of one of those parts?

A. 3.36

B. 1.36

C. 3.6

D. 0.36

44) Find the slope of the line.

A. $\frac{1}{3}$

B. $\frac{1}{6}$

C. 3

D. $-\frac{1}{6}$

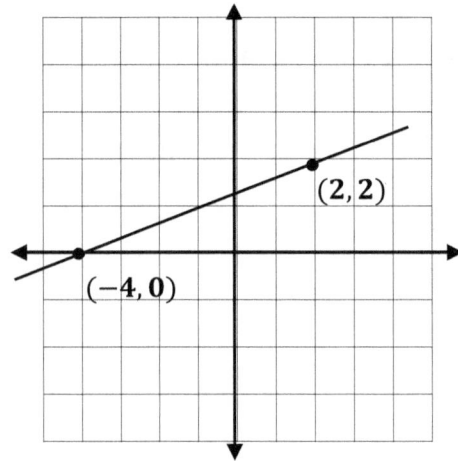

45) Amelia cuts a piece of birthday cake as shown below. What is the volume of

the piece of cake?

A. 560 cm^3

B. 280 cm^3

C. 140 cm^3

D. 880 cm^3

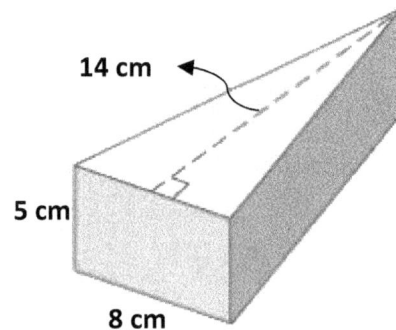

46) Let $f(x) = 3x - 6$. If $f(a) = -15$ and $f(b) = 6$, then what is $f(a + b)$?

A. -8

B. 12

C. -3

D. -15

"End of GED Mathematical Reasoning Practice Test 2."

Answers and Explanations

Answer Key

Now, it's time to review your results to see where you went wrong and what

areas you need to improve!

GED Math Practice Test

Practice Test 1

1	C	16	D	31	A
2	D	17	A	32	D
3	B	18	B	33	C
4	A	19	B	34	B
5	B	20	C	35	D
6	C	21	D	36	B
7	A	22	C	37	C
8	B	23	B	38	B
9	A	24	C	39	A
10	B	25	C	40	D
11	B	26	A	41	B
12	B	27	D	42	D
13	C	28	B	43	A
14	A	29	A	44	B
15	C	30	D	45	D
				46	C

Practice Test 2

1	C	16	C	31	C
2	B	17	B	32	B
3	A	18	C	33	D
4	C	19	A	34	D
5	D	20	D	35	C
6	A	21	C	36	A
7	C	22	D	37	B
8	B	23	A	38	D
9	D	24	C	39	B
10	C	25	D	40	D
11	C	26	B	41	C
12	D	27	A	42	D
13	A	28	D	43	C
14	C	29	A	44	A
15	D	30	C	45	B
				46	C

GED Practice Test 1

Answers and Explanations

1) Answer: C

$\frac{-29}{4} = -7.25$ and $\frac{23}{5} = 4.6$, then the odd numbers are:

$(-7, -5, -3, -1, 1, 3)$

2) Answer: D

The distance between two points always is positive. Use formula:

$AB = |b - a|$ or $|a - b|$

$|-4 - 12|$ or $|12 - (-4)| = |12 + 4|$

3) Answer: B

Sum of the measures of the angles of a triangle is 180, if two angles are 60 then the third one is 60, then the triangle is equilateral triangle, and all side are equal.

4) Answer: A

All factors of 20 are: 1, 2, 4, 5, 10, 20, then sum of them is 42.

5) Answer: B

$\frac{1}{3} + \frac{3}{5} = \frac{5}{15} + \frac{9}{15} = \frac{14}{15}$

$1 - \frac{14}{15} = \frac{15}{15} - \frac{14}{15} = \frac{1}{15}$

6) Answer: C

Each number is the sum of the previous and the number 2 places to the left.

Which is mean: $12 + 5 = 17$

7) Answer: A

Start to list factors of each number:

Factors of 12: 1, 2, 3, 4, 6, 12

Factors of 21: 1, 3, 7, 21

Factors of 32: 1, 2, 4, 8, 16, 32

The greatest common factor is 1.

8) Answer: B

5 more: +5

Ratio: ÷ ; Ratio of a number to 7: $\frac{x}{7}$

4 less: -4 ; 4 less than the number: $x - 4$

$5 + \frac{x}{7} = x - 4$

9) Answer: A

Rewriting each fraction with common denominator or converting each fraction to decimal and order the decimal from least to greatest.

$\frac{4}{9} = 0.44$ $\quad$ $\frac{7}{12} = 0.58$ $\quad$ $\frac{1}{5} = 0.2$ $\quad$ $\frac{14}{15} = 0.93$ $\quad$ $\frac{16}{21} = 0.76$

10) Answer: B

calculating Elena's total earnings:

35 hours × \$7.60 an hour = \$266

Next, divide this total by her brother's hourly rate:

\$266 ÷ \$13.30 = 20 hours

11) Answer: B

move decimal point in divisor so last digit is in the unit place (0.8 to 8)

move decimal point in dividend same number of places to the right,

(31.84 to 318.4)

divide (3,184 ÷ 8) = 398

insert a decimal point into the answer above the decimal point in the dividend

(39.8)

12) Answer: B

Circumference= $2\pi r = 2 \times \pi \times 14 = 28\pi$

13) Answer: C

Be careful with the conversion factor (per hundred gallons; NOT per gallon).

$$26{,}200 \times \frac{0.65}{100} = 170.3$$

$$170.3 + 4.80 = 175.1$$

14) Answer: A

$$\frac{(4x^5 - 8x^2)}{(2x^4 - 4x)} = \frac{4x^2(x^3 - 2)}{2x(x^3 - 2)} = 2x$$

15) Answer: C

For odd index we can have negative radicand.

In the even index, negative radicand is undefined.

$\sqrt{-81}$ has a negative number under the even index, so it is non-real,

Negative numbers don't have real square roots, because negative and positive

integer squared is either positive or 0.

16) Answer: D

$2f(4a) = 864 \rightarrow$ (divide by 2): $f(4a) = 432$

(subtitute 4a) $3(4a)^2 = 432 \rightarrow$ (divide by 3): $(4a)^2 = 144 \rightarrow 16a^2 =$

$144 \rightarrow a^2 = 9 \rightarrow a = 3$

17) Answer: A

2 hours and 45 minutes is 2.25 hour.

$R \times T = D \rightarrow 108 \times 2.25 = D \rightarrow D = 243$ miles

18) Answer: B

$14x - 8 \le 16x + 4 \rightarrow Add\ 8: 14x - 8 + 8 \le 16x + 4 + 8 \rightarrow$

$subtract\ 14x: 14x - 16x \le 12 \rightarrow -2x \le 12 \rightarrow x \ge -6$

19) Answer: B

Use formula to raise a number: $(x^a)^b = x^{ab}$

$3^8 = (3^2)^3 = 9^3$

20) Answer: C

Simple interest rate: I = prt (I = interest, p = principal, r = rate, t = time)

$$I = 3{,}800 \times 0.0275 \times 4 = 418$$

21) Answer: D

A straight angle is an angle measured exactly $180°$

$$70° + 38° = 108°$$

$$180° - 108° = 72°$$

22) Answer: C

$$\frac{56x^5 y^4 z^{-4}}{16\, x^2 y^6 z^3} = \frac{56}{16} \times \frac{x^5}{x^2} \times \frac{y^4}{y^6} \times \frac{z^{-4}}{z^3} = \frac{7}{2} \times x^3 \times \frac{1}{y^2} \times \frac{1}{z^7} = \frac{7x^3}{2\, y^2 z^7}$$

23) Answer: B

$$m = \frac{y_2 - y_1}{x_2 - x_1} = \frac{2 - (-2)}{7 - 4} = \frac{4}{3}$$

$$y - y_1 = m(x - x_1) \rightarrow y - (-2) = \frac{4}{3}(x - 4)$$

$$y + 2 = \frac{4}{3}(x - 4) \rightarrow 3(y + 2) = 4(x - 4) \rightarrow 3y + 6 = 4x - 16$$

$$3y - 4x = -16 - 6 \rightarrow 3y - 4x = -22$$

24) Answer: C

$$\frac{x^3 - 2}{y + 1} = \frac{(-2)^3 - 2}{4 + 1} = \frac{-10}{5} = -2$$

25) Answer: C

Ethan paid $3.8 for one hours and $0.65 for each of one fourth-hour period after that.

$$3 \times 0.65 = 1.95$$

$$3.8 + 1.95 = 5.75$$

26) Answer: A

Cross multiply and isolate x: $\frac{c+d}{3dx} = \frac{2}{7} \rightarrow 7(c + d) = 2(3dx) \rightarrow x = \frac{7(c+d)}{6d}$

$$x = \frac{7}{6}\left(\frac{c}{d} + \frac{d}{d}\right) = \frac{7}{6}\left(\frac{c}{d} + 1\right)$$

27) Answer: D

$$x^4 - 9x^2 = x^2(x^2 - 9) = x^2(x - 3)(x + 3)$$

28) Answer: B

Factoring: $(x - 6)(x - 7) = 0 \rightarrow \begin{cases} x - 6 = 0 \rightarrow x = 6 \\ x - 7 = 0 \rightarrow x = 7 \end{cases}$

29) Answer: A

Two points are $(0, 3)$ and $(8, 5)$

$$m = \frac{y_2 - y_1}{x_2 - x_1} = \frac{5 - 3}{8 - 0} = \frac{2}{8} = \frac{1}{4}$$

$$y - y_1 = m(x - x_1) \rightarrow y - 3 = \frac{1}{4}(x - 0)$$

$$y - 3 = \frac{1}{4}x \rightarrow y = \frac{1}{4}x + 3$$

30) Answer: D

$$\begin{cases} 3 \times (-2x + 4y = -6) \\ 4 \times (x - 3y = 5) \end{cases} \rightarrow \begin{cases} -6x + 12y = -18 \\ 4x - 12y = 20 \end{cases} \rightarrow \text{add two equations:}$$

$$-2x = 2 \rightarrow x = -1$$

31) Answer: A

Triangle third side rule: length of the one side of a triangle is less than the sum of the lengths of the other two sides and greater than the positive difference of the lengths of the other two sides.

the third side is less than $3 + 8 = 11$ and greater than $8 - 3 = 5$

32) Answer: D

Mean = $(101 + 113 + 128 + 132 + 153 + 159) \div 6$

$= 786 \div 6 = 131$

Only 141 can increase the mean.

33) Answer: C

The mode is the value which occurs with the greatest frequency. New York and Ohio are greatest and the same frequency (3 times).

34) Answer: B

convert the given distance, 9 km, into centimeters, (units on the map)

9 km = 9,000 m = 900,000 cm

divide by the ratio 1:30,000.

$\frac{900,000}{30,000} = 30$ cm

35) Answer: D

Area of square is: $a^2 = (\sqrt{5})^2 = 5$

36) Answer: B

$3x^3y^2 \, (2x^3y^4)^4 == 3x^3y^2(16x^{12}y^{16}) = 48x^{15}y^{18}$

37) Answer: C

Change percent to decimal: $20\% = 0.20$

$0.20 \times 150 = 30$

38) Answer: B

Baseballs and basketballs are spherical. The volume of sphere: $v = \frac{4}{3}\pi r^3$

$36\pi = \frac{4}{3}\pi r^3 \rightarrow 36 = \frac{4}{3}r^3 \rightarrow 108 = 4r^3 \rightarrow r^3 = 27 \rightarrow r = 3$

$d = 2r \rightarrow d = 2 \times 3 = 6$

39) Answer: A

Use percent formula: Part $= \frac{\text{percent}\times\text{whole}}{100}$

$154 = \frac{\text{percent} \times 140}{100} \Rightarrow \frac{154}{1} = \frac{\text{percent} \times 140}{100}$, cross multiply.

$15,400 = \text{percent} \times 140$, divide both sides by 140. $\rightarrow$ percent $= 110$

40) Answer: D

The sample shows that 4 out of 600 phones will be faulty. Consequently, a proportion can be set up.

$\frac{4}{600} = \frac{P}{24,000}$ (P: Faulty Phone) $\rightarrow P = \frac{4\times24,000}{600} = 160$

41) Answer: B

We can Write an equation to solve the problem.

Emma Books =Mia books +4 →Mia book = Emma – 4=b – 4

Emma + Mia=16

$b + b - 4 = 16 \rightarrow 2b - 4 = 16 \rightarrow 2b = 16 + 4 \rightarrow b = 10$

42) Answer: D

$\text{Probability} = \dfrac{number\ of\ desired\ outcomes}{number\ of\ total\ outcomes} = \dfrac{5}{7+5+13} = \dfrac{5}{25} = \dfrac{1}{5} = 0.20$

43) Answer: A

Surface Area of a cylinder = 2πr (r + h),

The radius of the cylinder is 3 inches, and its height are 9 inches. π is 3.14. Then:

Surface Area of a cylinder = 2 (3.14) (3) (3 + 9) = 226.08 inches

44) Answer: B

use the Pythagorean theorem to find the value of unknown side.

$a^2 + b^2 = c^2 \rightarrow c^2 = 10^2 + 24^2 \rightarrow a^2 = 100 + 576 = 676 \rightarrow c = 26$

45) Answer: D

A line perpendicular to a line with slope m has a slope of $-\dfrac{1}{m}$.

So, the slope of the line perpendicular to the given line is $-\dfrac{1}{\frac{a}{b}} = -\dfrac{b}{a}$.

46) Answer: C

If 24% of the total number of customers is female, then 100% – 24% = 76% of

the customers are male. Calculate 76% of the total.

0.76× 750 = 570.

GED Practice Test 2

Answers and Explanations

1) Answer: C

The smallest prime number is 2, and the largest even negative integer is −2.

$2 + 5(−2) = 2 − 10 = −8$.

2) Answer: B

The difference between his income and his cost is monthly saving.

January: $5,145− $3,990= $1,155

February: $5,620− $3,885= $1,735

March: $5,380− $3,780= $1,600

3) Answer: A

$8\frac{5}{18} - 5\frac{1}{6} + 2\frac{2}{3} = (8 − 5 + 2)\frac{5}{18} − \frac{3}{18} + \frac{12}{18} = 5(\frac{2}{18} + \frac{12}{18}) = 5\frac{14}{18} = 5\frac{7}{9}$

4) Answer: C

$3^4 + 3^3 + 3 = 81 + 27 + 3 = 111$

5) Answer: D

$\frac{180}{450} = \frac{180}{45} \times \frac{1}{10} = 4 \times \frac{1}{10} = 0.4$

6) Answer: A

$|2x − 4| = 8 \rightarrow \begin{cases} 2x − 4 = 8 \rightarrow 2x = 12 \rightarrow x = 6 \\ 2x − 4 = −8 \rightarrow 2x = −4 \rightarrow x = −2 \end{cases}$

7) Answer: C

Multiply equation (2) by 5. Add two equations [(1) +5(2)]:

$\begin{cases} 3x − 5y = 6 \\ 10x + 5y = 20 \end{cases} \rightarrow 13x = 26 \rightarrow x = 2$

Substitute $x = 2$ into equation (1): $3(2) − 5y = 6 \rightarrow −5y = 0 \rightarrow y = 0$

8) Answer: B

2 hours equals 120 minutes.

The train passes 5 stations every 24 minutes: $\frac{5}{24}$

$\frac{5}{24} = \frac{x}{120} \rightarrow x = \frac{120 \times 5}{24} = 25$ stations

9) Answer: D

Difference in percent: $15\% - 13\% = 2\%$

$2\% \times 130 = 0.02 \times 130 = 2.6$

10) Answer: C.

$C = \pi d = 2\pi r = 2\pi \times 27 = 54\,\pi$

11) Answer: C

$V = l \times w \times h = 16 \times 11 \times 6 = 1,056\ in^3$

12) Answer: D

$\sqrt[5]{2^{-10}} = \sqrt[5]{\frac{1}{2^{10}}} = \frac{\sqrt[5]{1}}{\sqrt[5]{2^{10}}} = \frac{1}{2^{\left(\frac{10}{5}\right)}} = \frac{1}{2^2} = 2^{-2} = \frac{1}{4}$

13) Answer: A

There are no values of the variable that make the equation true.

14) Answer: C

Volume of cylinder: $V = \pi r^2 h = \pi \times 5^2 \times 12 = 300\pi$

Volume of cone: $V = \frac{1}{3}\pi r^2 h = \frac{1}{3}\pi \times 5^2 \times 6 = 50\pi$

$300\pi + 50\pi = 350\pi.$

15) Answer: D

$9.23 \times 10^{-6} = 0.00000923$

16) Answer: C

$x + 50 + 33 + 47 = 180 \rightarrow x + 130 = 180 \rightarrow x = 50$

17) Answer: B

use the Pythagorean theorem to find the value of unknown side.

$a^2 + b^2 = c^2 \rightarrow 17^2 = a^2 + 15^2 \rightarrow a^2 = 289 - 225 = 64 \rightarrow a = 8$

18) Answer: C

Use percent formula: $\text{Part} = \frac{\text{percent} \times \text{whole}}{100}$

Part$= \frac{40 \times 420}{100} = 168$

Last price: $420 + 168 = \$588$

19) Answer: A

Area of trapezoid: $\frac{(a+b)}{2} \times h \rightarrow A = \frac{(14+16)}{2} \times 12 = 180$

Area of triangle: $\frac{b \times h}{2} \rightarrow A = \frac{8 \times 12}{2} = 48$

Area of shaded region: $180 - 48 = 132$

20) Answer: D

$xy - 6x = 64 \rightarrow x(y - 6) = 64$

$8x = 64 \rightarrow x = 8$

21) Answer: C

Multiply both sides by 3: $3 \times \left(\frac{x}{3}\right) = 3 \times (x - 8) \rightarrow x = 3x - 24$

Subtract both side by x and add 24 to both sides: $24 + x - x = 3x - x - 24 +$

$24 \rightarrow 2x = 24 \rightarrow x = 12$

22) Answer: D

$\begin{cases} 3 \times (-3x - 2y = 5) \\ -2 \times (-4x - 3y = 7) \end{cases} \rightarrow \begin{cases} -9x - 6y = 15 \\ 8x + 6y = -14 \end{cases} \rightarrow$add two equations:

$-x = 15 - 14 \rightarrow x = -1$

23) Answer: A

There are 3 parts labeled "H" out of a total of 8 equal parts.

The probability of not spinning at "H" is 5 out of 8.

24) Answer: C

The best estimate of the product of 21 and 2.06 to the nearest whole number.

Since 21 is already a whole number and 2.06 is closer to 2 than 3. Option C is the correct answer.

25) Answer: D

Fuel consumption rate $= \frac{12}{60} = 0.2$ liter per mile

Cost: $1,500 \times 0.2 \times 2.05 = 615$

26) Answer: B

Point $1(x_A, y_A) = (-4, -5)$

Point $2(x_B, y_B) = (8, 4)$

Distance between two points $= \sqrt{(x_B - x_A)^2 + (y_B - y_A)^2}$

$$\rightarrow d = \sqrt{\left(8 - (-4)\right)^2 + \left(4 - (-5)\right)^2} = \sqrt{12^2 + 9^2} = \sqrt{144 + 81}$$

$$\rightarrow d = \sqrt{225} = 15$$

27) Answer: A

Ordered values: 47, 83, 98, 152, 158, 224, 402.

The 4th number is median.

28) Answer: D

$\sqrt{5x - 4} = 4 \rightarrow 5x - 4 = 16 \rightarrow 5x = 20 \rightarrow x = 4$

29) Answer: A

Get the equation into slop intercept form:

$y = mx + b$ where m is slope and b is y-intercept.

$3x = 8y - 24$; Add 24 to both sides $\rightarrow 3x + 24 = 8y$; Dividing by 8:

$\rightarrow y = \frac{3}{8}x + 3$ Thus, the y-intercept is 3.

30) Answer: C

$3^{2x} = 729 \rightarrow (3^2)^x = 729 \rightarrow 3^{2x} = 3^6 \rightarrow 2x = 6 \rightarrow x = 3$

31) Answer: C

$15: 3 \times 5$

$30: 2 \times 3 \times 5$

LCM $(15, 30) = 2 \times 3 \times 5 = 30$

32) Answer: B

Surface area (SA): $6a^2 \to 24 = 6a^2 \to a^2 = 4 \to a = 2$ cm

Volume (v): $a^3 \to v = 2^3 = 8$ cm^3

33) Answer: D

Write the equation into standard form: $x^2 + 5x = 36 \to x^2 + 5x - 36 = 0$

Factor this expression: $(x - 4)(x + 9) = 0 \to x = 4 \ or, x = -9 \to x^2 = \begin{cases} 16 \\ 81 \end{cases}$

34) Answer: D

The minimum amount of water: $9 \times \frac{4}{9} = 4; \ 4 \div 4 = 1$

The maximum amount of water: $9 \times \frac{4}{9} = 4$

Amount of water in all pitchers: $1 < w < 4$

35) Answer: C

The range is: highest value – lowest value.

R $= 52 - 35 = 17$

36) Answer: A

d $= 2$r $\to$ r $= \frac{d}{2} = \frac{8}{2} = 4$

Area of circle (A): $\pi r^2 = \pi \times 4^2 = 50.24 \simeq 50.2$

37) Answer: B

Factor the expression: $\frac{2x^2 - 11x - 6}{4(x^2 - \frac{1}{4})} = \frac{(x-6)(2x+1)}{4x^2 - 1} = \frac{(x-6)(2x+1)}{(2x-1)(2x+1)} = \frac{x-6}{2x-1}$

38) Answer: D

$(\frac{1}{3})^{-4} = \frac{1}{3^{-4}} = 3^4 = 81$

39) Answer: B

$$\frac{(2x^{-2}y^3)^4}{48y^{-5}z^{-4}} = \frac{16x^{-8}y^{12}}{48y^{-5}z^{-4}} = \frac{y^{17}z^4}{3x^8}$$

40) Answer: D

Rate: $\frac{\$1}{1Kg} = \frac{100¢}{1,000g} \rightarrow 1^{\$}/_{kg} = 0.1^{¢}/_{g}$

The price is: $30 \times 0.1 = 3$cent per kilogram.

41) Answer: C

$3P = 9 \rightarrow P = 3$ The price of 2012 is three times of the price of 2006.

42) Answer: D

First equation: $4y - 5 = 12x + 11 \rightarrow 4y = 12x + 16 \rightarrow y = 3x + 4 \rightarrow m_1 = 3$

Second question: $3y - 2 = x + 6 \rightarrow 3y = x + 8 \rightarrow y = \frac{1}{3}x + \frac{8}{3} \rightarrow m_2 = \frac{1}{3}$

$m_1 = 3$ and $m_2 = \frac{1}{3}$, they aren't equal slopes or negative reciprocals.

43) Answer: C

Area $= 4\frac{1}{2} \times 3\frac{1}{5} = \frac{9}{2} \times \frac{16}{5} = \frac{144}{10} = \frac{72}{5}$ square feet.

One-fifth $= \frac{72}{5} \div 4 = \frac{72}{5} \times \frac{1}{4} = \frac{72}{20} = \frac{36}{10} = 3.6$

44) Answer: A

Two points are $(2, 2)$ and $(-4,0) \rightarrow m = \frac{y_2 - y_1}{x_2 - x_1} = \frac{0-2}{-4-2} = \frac{-2}{-6} = \frac{1}{3}$

45) Answer: B

$V = \left(\frac{base \times height}{2}\right) \times$ height of prism $\rightarrow V = \left(\frac{14 \times 8}{2}\right) \times 5 = 280$

46) Answer: C

$f(a) = 3a - 6$ and $f(a) = -15$: $3a - 6 = -15$ so that $3a = -9$ and $a = -3$.

$f(b) = 3b - 6$ and $f(b) = 6$: $3b - 6 = 6$ so that $3b = 12$ and $b = 4$

Finally, $f(a + b) = f(-3 + 4) = f(1)$

$f(1) = 3(1) - 6 = -3$.

"END"

www.ingramcontent.com/pod-product-compliance
Lightning Source LLC
La Vergne TN
LVHW081315060426
835509LV00015B/1525